SPACE SCIENCE IN THE TWENTY-FIRST CENTURY: IMPERATIVES FOR THE DECADES 1995 TO 2015

PLANETARY AND LUNAR EXPLORATION

Task Group on Planetary and Lunar Exploration
Space Science Board
Commission on Physical Sciences, Mathematics, and Resources
National Research Council

NATIONAL ACADEMY PRESS
Washington, D.C. 1988

National Academy Press • 2101 Constitution Avenue, N.W. • Washington, D. C. 20418

NOTICE: The project that is the subject of this report was approved by the Governing Board of the National Research Council, whose members are drawn from the councils of the National Academy of Sciences, the National Academy of Engineering, and the Institute of Medicine. The members of the committee responsible for the report were chosen for their special competences and with regard for appropriate balance.

This report has been reviewed by a group other than the authors according to procedures approved by a Report Review Committee consisting of members of the National Academy of Sciences, the National Academy of Engineering, and the Institute of Medicine.

The National Academy of Sciences is a private, nonprofit, self-perpetuating society of distinguished scholars engaged in scientific and engineering research, dedicated to the furtherance of science and technology and to their use for the general welfare. Upon the authority of the charter granted to it by the Congress in 1863, the Academy has a mandate that requires it to advise the federal government on scientific and technical matters. Dr. Frank Press is president of the National Academy of Sciences.

The National Academy of Engineering was established in 1964, under the charter of the National Academy of Sciences, as a parallel organization of outstanding engineers. It is autonomous in its administration and in the selection of its members, sharing with the National Academy of Sciences the responsibility for advising the federal government. The National Academy of Engineering also sponsors engineering programs aimed at meeting national needs, encourages education and research, and recognizes the superior achievements of engineers. Dr. Robert M. White is president of the National Academy of Engineering.

The Institute of Medicine was established in 1970 by the National Academy of Sciences to secure the services of eminent members of appropriate professions in the examination of policy matters pertaining to the health of the public. The Institute acts under the responsibility given to the National Academy of Sciences by its congressional charter to be an adviser to the federal government and, upon its own initiative, to identify issues of medical care, research, and education. Dr. Samuel O. Thier is president of the Institute of Medicine.

The National Research Council was organized by the National Academy of Sciences in 1916 to associate the broad community of science and technology with the Academy's purposes of furthering knowledge and advising the federal government. Functioning in accordance with general policies determined by the Academy, the Council has become the principal operating agency of both the National Academy of Sciences and the National Academy of Engineering in providing services to the government, the public, and the scientific and engineering communities. The Council is administered jointly by both Academies and the Institute of Medicine. Dr. Frank Press and Dr. Robert M. White are chairman and vice chairman, respectively, of the National Research Council.

Support for this project was provided by Contract NASW 3482 between the National Academy of Sciences and the National Aeronautics and Space Administration.

Library of Congress Catalog Card Number 87-43335

ISBN 0-309-03885-5

Printed in the United States of America

TASK GROUP ON
PLANETARY AND LUNAR EXPLORATION

Donald M. Hunten, University of Arizona, Chairman
Arden L. Albee, California Institute of Technology
David C. Black, NASA Headquarters
Jacques Blamont, C.N.E.S.
William V. Boynton, University of Arizona
Robert A. Brown, Space Telescope Science Institute
A.G.W. Cameron, Center for Astrophysics
Thomas M. Donahue, University of Michigan
Larry W. Esposito, University of Colorado
Ronald Greeley, Arizona State University
Eugene H. Levy, University of Arizona
Harold Masursky, U.S. Geological Survey
David D. Morrison, University of Hawaii at Manoa
George W. Wetherill, Carnegie Institution of Washington

Paul F. Uhlir, *Staff Officer*
Anne L. Pond, *Secretary*

STEERING GROUP

Thomas M. Donahue, University of Michigan, Chairman
Don L. Anderson, California Institute of Technology
D. James Baker, Joint Oceanographic Institutions, Inc.
Robert W. Berliner, Pew Scholars Program, Yale University
Bernard F. Burke, Massachusetts Institute of Technology
A. G. W. Cameron, Harvard College Observatory
George B. Field, Center for Astrophysics, Harvard University
Herbert Friedman, Naval Research Laboratory
Donald M. Hunten, University of Arizona
Francis S. Johnson, University of Texas at Dallas
Robert Kretsinger, University of Virginia
Stamatios M. Krimigis, Applied Physics Laboratory
Eugene H. Levy, University of Arizona
Frank B. McDonald, NASA Headquarters
John E. Naugle, Chevy Chase, Maryland
Joseph M. Reynolds, The Louisiana State University
Frederick L. Scarf, TRW Systems Park
Scott N. Swisher, Michigan State University
David A. Usher, Cornell University
James A. Van Allen, University of Iowa
Rainer Weiss, Massachusetts Institute of Technology

Dean P. Kastel, *Study Director*
Ceres M. Rangos, *Secretary*

SPACE SCIENCE BOARD

Thomas M. Donahue, University of Michigan, Chairman
Philip H. Abelson, American Association for the Advancement of Science
Roger D. Blandford, California Institute of Technology
Larry W. Esposito, University of Colorado
Jonathan E. Grindlay, Center for Astrophysics
Donald N. B. Hall, University of Hawaii
Andrew P. Ingersoll, California Institute of Technology
William M. Kaula, NOAA
Harold P. Klein, The University of Santa Clara
John W. Leibacher, National Solar Observatory
Michael Mendillo, Boston University
Robert O. Pepin, University of Minnesota
Roger J. Phillips, Southern Methodist University
David M. Raup, University of Chicago
Christopher T. Russell, University of California, Los Angeles
Blair D. Savage, University of Wisconsin
John A. Simpson, Enrico Fermi Institute, University of Chicago
George L. Siscoe, University of California, Los Angeles
L. Dennis Smith, Purdue University
Darrell F. Strobel, Johns Hopkins University
Byron D. Tapley, University of Texas at Austin

Dean P. Kastel, *Staff Director*
Ceres M. Rangos, *Secretary*

COMMISSION ON PHYSICAL SCIENCES, MATHEMATICS, AND RESOURCES

Norman Hackerman, Robert A. Welch Foundation, *Chairman*
George F. Carrier, Harvard University
Dean E. Eastman, IBM Corporation
Marye Anne Fox, University of Texas
Gerhart Friedlander, Brookhaven National Laboratory
Lawrence W. Funkhouser, Chevron Corporation (retired)
Phillip A. Griffiths, Duke University
J. Ross Macdonald, University of North Carolina, Chapel Hill
Charles J. Mankin, Oklahoma Geological Survey
Perry L. McCarty, Stanford University
Jack E. Oliver, Cornell University
Jeremiah P. Ostriker, Princeton University Observatory
William D. Phillips, Mallinckrodt, Inc.
Denis J. Prager, MacArthur Foundation
David M. Raup, University of Chicago
Richard J. Reed, University of Washington
Robert E. Sievers, University of Colorado
Larry L. Smarr, National Center for Supercomputing Applications
Edward C. Stone, Jr., California Institute of Technology
Karl K. Turekian, Yale University
George W. Wetherill, Carnegie Institution of Washington
Irving Wladawsky-Berger, IBM Corporation

Raphael G. Kasper, *Executive Director*
Lawrence E. McCray, *Associate Executive Director*

Foreword

Early in 1984, NASA asked the Space Science Board to undertake a study to determine the principal scientific issues that the disciplines of space science would face during the period from about 1995 to 2015. This request was made partly because NASA expected the Space Station to become available at the beginning of this period, and partly because the missions needed to implement research strategies previously developed by the various committees of the board should have been launched or their development under way by that time. A two-year study was called for. To carry out the study the board put together task groups in earth sciences, planetary and lunar exploration, solar and space physics, astronomy and astrophysics, fundamental physics and chemistry (relativistic gravitation and microgravity sciences), and life sciences. Responsibility for the study was vested in a steering group whose members consisted of the task group chairmen plus other senior representatives of the space science disciplines. To the board's good fortune, distinguished scientists from many countries other than the United States participated in this study.

The findings of the study are published in seven volumes: six task group reports, of which this volume is one, and an overview report of the steering group. I commend this and all the other task group reports to the reader for an understanding of the challenges that confront the space sciences and the insights they promise for the next century. The official recommendations of the study are those to be found in the steering group's overview.

        Thomas M. Donahue, Chairman
        Space Science Board

Contents

1. INTRODUCTION   1
2. SCIENTIFIC GOALS AND RECOMMENDATIONS   4
   Goals of Planetary Exploration, 4
   A Balanced Planetary Program, 7
   A Mars Focus, 8
   Summary of Recommendations, 9
      Scientific Investigations, 10
      Technical Developments, 17
      International Collaboration, 18
3. STATUS OF PLANETARY SCIENCE IN 1995   19
   Overview, 19
      State of Planetary Exploration as of 1995, 19
      Scientific Questions as of 1995, 23
   Planetary Geosciences, 25
      Scientific Objectives for Planetary Geosciences, 26
      The Inner Solar System, 28
      Rocky Satellites, 37
   Atmospheres, 41
      Earth, Mars, and Venus, 41
      Titan, 44
      Io and the Plasma Torus, 46

Jovian Planets, 47
Rings, 49
Interiors of the Giant Planets, 53
Planetary Magnetism, 57
    Generation of Planetary Magnetic Fields, 58
    Earth's Magnetic Field, 61
    Studying Planetary Magnetic Fields, 62
Primitive Bodies and the Origin of the Solar System, 64
    The Origin of the Sun and Planets, 64
    Search for and Study of Other Planetary Systems, 70
    Asteroids, Small Satellites, and Meteorites, 73
Comets, 80
    General Characteristics, 80
    State of Knowledge in 1995, 82

4. **FUTURE PROGRAMS**   88

Proposed Missions, 88
    Programs for Planetary Geosciences, 88
    Programs for the Outer Solar System, 94
    Future Missions and Programs for Primitive Bodies and the Origin of the Solar System, 96
A Program for Intensive Exploration of Mars, 99
    Why Mars?, 99
    Scientific Objectives for a Mars Focus, 102
    Role of Humans in Intensive Mars Exploration, 104
    A Phased Approach, 105
Recommendations, 106
    Exploration of the Solar System, 106
    Exploration of Mars, 111

1
Introduction

The solar system consists of nine known planets orbiting the Sun, along with a larger number of moons, asteroids, planetary rings, and comets. Among the mysteries that have occupied human thought throughout history are the mechanisms by which the solar system came into existence, the laws and physical processes that shape the evolution and behavior of planets, and the relationship of the solar system to the wider cosmos. These questions continue to stimulate modern planetary science.

In addition to illuminating some of these long-standing scientific questions, planetary studies have additional significance from a human perspective. Planets are likely to be the only bodies in the universe capable of supporting advanced life. Among its other objectives, planetary science seeks to understand the formation of life-supporting planets and the conditions under which life arises and develops. The answers to these questions promise to help shape human perceptions about our origins and about our place in the universe.

The Sun and planets occupy a flattened, disk-shaped region of space extending some 40 times farther from the Sun than does the orbit of Earth. All of the planets go around the Sun in nearly circular orbits, in the same direction, and nearly in the same plane. The planets near the Sun—known collectively as the terrestrial planets—are comparable in size to Earth; like Earth, the terrestrial planets are composed largely of rock and metal. Most have atmospheres; some have (or once had) hydrospheres. The planets in the outer solar system—known as the jovian planets—are much larger than the terrestrial planets and composed largely of gas and ice. Many of their satellites resemble terrestrial planets, but with a much larger endowment of low-temperature and icy material.

According to our present understanding, the Sun and planets formed together, some 4.5 billion years ago, from a flattened nebula of interstellar gas and dust. The overall pattern of the solar system, including the arrangement of planetary orbits and the variation of planetary mass and composition with distance from the Sun, is thought to have arisen as a natural consequence of the structure and evolution of the protoplanetary disk. The phenomena that led to the formation of this planetary system are thought to be representative of the processes prevailing in star formation. This suggests that many stars may have similar planetary systems associated with them.

The planets have evolved considerably since their formation. Large sources of internal heat—trapped gravitational energy and energy liberated by radioactive decay—have driven a continuing evolution in most of the planetary bodies. Tidal energy is important for some satellites. As a consequence, many of the planets have thoroughly differentiated, their heavy constituents having sunk to their centers. To this day, the continuing escape of heat from planetary interiors drives convection, which is responsible for the observed dynamical behaviors of planetary surfaces and interiors. On Earth, volcanoes, earthquakes, sea floor spreading, and continental drift are contemporary consequences of Earth's thermally agitated interior.

Through a combination of gas trapped at the time of planet formation and volatiles released during subsequent planetary evolution, most of the planets have developed atmospheres. On Earth, the water-rich atmosphere has played a major role in the formation of terrestrial life—still the only known instance of this remarkable phenomenon—which has existed for more than 3 billion years. During that time it has developed into an extremely complicated system. Many of the characteristics of Earth's surface and atmosphere result from the effects of biological activity. Conversely, Earth's atmosphere and hydrosphere remain dominating factors in the existence and continuing success of life on Earth.

The extreme evolutionary processes through which the major planets have passed have resulted in the obliteration of most of the primordial characteristics of planetary bodies and of their constituent material. Thus, from the planets themselves we can glean few clues about the detailed processes of their origin. What information the planets do communicate about the solar system's origin is contained in their orbits, their masses, their bulk compositions, their rings, and their satellite systems.

However, the solar system contains many small objects that have not suffered such severe processing since their formation. Comets are believed to be relatively well-preserved accumulations of matter left from the time of the solar system's birth. Comets are rich in volatile matter, suggesting that they formed in the relatively cold outer reaches of the early solar system. Asteroids seem to have formed closer to the Sun than did the comets—in the general vicinity of their present locations, mostly between Mars and Jupiter. Because most asteroids are small, they have not undergone continuing evolution in the manner of planets, although they do show evidence of early evolutionary processing. Thus asteroids are thought to preserve evidence of processes, and perhaps sources of heat, that are no longer active in the solar system.

The solar system is the largest region of space available to humans for in situ measurement and experimentation. Much of our understanding of more remote astrophysical systems is based on the knowledge gained from investigation of solar system objects, which has revealed the actual behavior of the physical laws in complex, large-scale systems.

2
Scientific Goals and Recommendations

GOALS OF PLANETARY EXPLORATION

The scientific goals motivating planetary exploration are:
- To understand the origin of the solar system.
- To understand the evolution and present states of the planets, including Earth.
- To learn what conditions lead to the origin of life, and how life modifies its environment.
- To learn how physical laws work in large systems.

To understand the origin of the solar system. Research aimed at understanding the origin of the solar system focuses on those objects thought to retain clues about the primordial conditions and processes that contributed to the system's formation. Although critical information is obtained from measurements made of the evolved planets, the most detailed clues come from investigations of those small primitive objects that have changed little since the time of formation in the protoplanetary nebula: comets, asteroids, and certain meteorites.

The cold, volatile-rich matter of comets is thought to contain the most faithfully preserved samples of condensed protoplanetary material remaining in the solar system. The asteroids form an ordered assemblage of protoplanetary fragments, which seem to remain near the original locations of their formation. The compositional and structural variations of objects in the asteroid belt are thought to reflect the radial variation of conditions in the protoplanetary nebula. Ongoing laboratory analyses of meteorites, which are fragments of asteroids and comets, already show the importance of the information that these objects can provide. Detailed study of comets and asteroids is expected to result in fundamental advances in our understanding of the solar system's formation.

Because the formation of our planetary system is thought to have been a natural consequence of the Sun's formation, without the intervention of special circumstances, and because the formation of our Sun is thought to have involved processes typical of star formation in general, planetary systems are believed to occur commonly in the universe, although none has yet been detected. Indeed, our present ideas lead so immediately to the conclusion that planetary systems occur frequently around stars that failure to find such systems would force a fundamental revision of our theories about the origin of our planetary system and about star formation. Important advances in our understanding of the formation of the solar system and in our understanding of planetary systems as a class are expected to come from studies of star-forming regions and from the discovery and study of other planetary systems.

To understand the evolution of the planets. Research aimed at understanding the evolution of the planets and the physical processes that govern their behavior and their environments concentrates on those bodies that exhibit most clearly the consequences of planetary evolution. Because we live on Earth, a terrestrial planet, the evolution and environment of terrestrial planets is of special interest. Substantial advances in understanding can be realized by investigating, as a class, the terrestrial planets and their close analogs.

The major targets of comparative terrestrial planet research beyond Earth are Mercury, Venus, Mars, and the Moon. In addition, studies of many of the outer-planet satellites and of the largest asteroids are expected to reveal important information about solid planet evolution.

Terrestrial planet research exploits the close relationship between behaviors and physical processes occurring on Earth and those occurring on many other planets. Indeed, much of what we can understand about the terrestrial planets derives from ideas and concepts that originated in studies of Earth. Conversely, planetary investigations of objects that evolved under conditions far different from those on Earth provoke us to achieve a deeper and more general grasp of natural terrestrial phenomena, as well as a more confident understanding of Earth's history. By exposing circumstances in which concepts based on terrestrial analogs fail, planetary investigations help us define the limits of applicability of these earth-based ideas.

To learn what conditions lead to the origin of life. Earth remains the only place where we know life has arisen and continued to flourish. Our search to understand the origin of life involves several planetary questions. It is important to know the physical conditions and chemical composition under which biological activity arose. It is also important to ascertain whether life forms, complex or incipient, have arisen elsewhere where they can be studied.

Presumably, life arose out of an organic, prebiotic medium and was preceded by an interval of chemical evolution, which led more or less continuously into biological evolution. By understanding the formation of the planets we will gain knowledge of the circumstances under which life arose on Earth. Many objects in the solar system seem not to have undergone substantial evolution since their formation. Some—Saturn's moon Titan, for example—are expected to carry important clues about the early material in which biological activity arose. These and other objects in the solar system—including Mars—may have supported prebiotic or early biotic evolution, leaving evidence that can be found today.

Investigations of the composition of cosmic matter and primitive solar system matter show that the basic building blocks of terrestrial life, including amino acids, occur naturally—at least in trace amounts. One of the most significant challenges in understanding the origin and distribution of life is to determine the extent to which special terrestrial conditions were involved in prebiotic chemical evolution. Detailed chemical assays of comets, asteroids, and other primitive objects will reveal the extent to which life could have arisen directly from preplanetary matter without an interval of special processing to condition the chemical mix.

To learn how physical laws work in large systems. Various phenomena result uniquely from the large scale of natural systems or from the long time scales over which slow processes work. Because these phenomena do not occur under normal laboratory conditions, it is only through direct observations in the solar system that we can expect to understand such important processes as planetary tectonism and volcanism, and cosmic plasma processes.

In situ studies of physical processes occurring in the solar system strengthen our overall understanding of the behavior of natural systems. Our ideas about observable phenomena throughout the universe are shaped by detailed solar system investigations. There is little prospect that such intense scrutiny will ever be extended to the more distant reaches of the universe. Thus, detailed investigations within the solar system will continue to be the foundation upon which is built most of our understanding of natural phenomena throughout the universe.

Investigations of large-scale physical processes encompass virtually all of the objects in the solar system. The giant planets provide clues about properties of matter under high pressures; planetary interiors and magnetospheres demonstrate the curious behaviors of magnetized fluids and plasmas; and planetary atmospheres and surfaces present puzzles about long-term evolution of the complex interacting systems that constitute planetary environments and interiors.

A BALANCED PLANETARY PROGRAM

Altogether, progress toward realizing these goals requires a balanced program of science and exploration encompassing studies of the planets and their satellites, asteroids and comets in our solar system, and star-forming regions and other planetary systems.

The history of science shows us that natural objects sometimes behave in astonishing ways. All scientific investigations begin with exploratory studies to establish the basic characteristics of the systems of interest and to discover their physical behaviors. This exploratory phase has been accomplished for part of the solar system, although significant work remains to be done.

Once the basic characteristics and behaviors are discovered, then investigations proceed on to intensive studies aimed at achieving a deeper understanding. Only very limited intensive studies of the solar system will have been accomplished by 1995. These will involve the Moon, Venus, Mars, and Jupiter. However, the Task Group on Planetary and Lunar Exploration envisions that in the years after 1995, planetary exploration will shift increasingly toward such intensive studies.

A variety of scientific techniques can be used to pursue the goals of planetary exploration. These include laboratory experimentation and theoretical analysis, astronomical observations, and in situ investigations using spacecraft. During the past 25 years the primary cause of the enormous advance in our understanding of the solar system has been the information obtained from scientific spacecraft. The task group envisions that such spacecraft investigations will continue to play the primary role in advancing our understanding of the solar system.

A MARS FOCUS

Planets and their environments exhibit extraordinary behaviors that, for fundamental reasons, cannot be predicted from first principles. The complexity and nonlinearity of planetary environments are such that a planet can, in principle, exist in a large variety of states with the same conditions imposed from outside. The possibility of living systems adds further complication to the variety of states in which a planet can exist. One of the major challenges of planetary science is to understand the evolution of terrestrial planets and to discover the possible varieties and causes of diverse planetary environments. Achievement of this goal requires comparative studies of the various terrestrial planets and satellites, as well as intensive studies of the changes that individual planetary environments undergo.

Of particular interest in the comparison of terrestrial planets is the puzzle posed by the triad of planets with atmospheres: Venus, Earth, and Mars. These three planets exhibit differences in their present environments and in their styles of evolution that seem large in comparison with the differences in their sizes, locations, and overall compositions. Solving this puzzle is important to us because the differences between these planets occur in those aspects of their environments most important for the viability of life.

Several ongoing and planned missions in the planetary program are directed toward the study of terrestrial planets. However, a number of conderations suggest that it may be appropriate to undertake a special focus on terrestrial planet science within the context of an overall balanced program, with Mars as the center of that focus.

Spacecraft investigations of Mars during the past 15 years reveal that that planet has undergone perplexing changes throughout its history. Although today the planet appears dry and cold, there is clear evidence of abundant flow of water several times in the past. Changes in the martian surface environment directly pertain to concerns about the behavior of Earth's environment. The prior presence of water on Mars raises important questions about its early, if temporary, suitability for life.

Photographs returned to Earth from Mariner and Viking spacecraft reveal spectacular geological formations and deep-cut relief. It is evident that detailed study of the martian crust will yield information about the character of the planet's past environments, their arrangement in time, and perhaps clues regarding the influences that produced such marked environmental change. These studies may even help us to understand the Earth's ice ages.

Of all the planets beyond Earth, Mars is the one most accessible to detailed study. It is relatively easy to reach with scientific spacecraft, and the surface is the most conducive to sustained operation of scientific instruments on mobile platforms. Furthermore, it is the only planet outside of the Earth-Moon system that we can currently consider for manned exploration and settlement.

SUMMARY OF RECOMMENDATIONS

In the next 30 years the task group expects to see an increase in our technical infrastructure in space. This increase should include both earth-orbital facilities for scientific investigations, and advanced capabilities for deep-space operation of scientific spacecraft. The recommendations put forward here will take advantage of our expanding capabilities in space. If implemented, the recommendations will advance our understanding of the solar system on the broad front that is needed to progress toward answering some of mankind's longest-standing questions about the cosmos. A recommended Mars focus within that broad-based program will further our understanding of the terrestrial planets, including Earth, and will address pressing questions about planetary environments and their stability.

The goals of planetary exploration are achieved primarily

through spacecraft missions, although various earth-based and orbital activities also serve an integral role. Below, the task group has outlined the activities and missions recommended for the period 1995 to 2015. These are also summarized in Table 2.1. Figures 2.1 through 2.4, which are in time sequence, illustrate the current and projected status in terms of actual missions.

This recommended plan is fully consistent with the reports of the Space Science Board's Committee on Planetary and Lunar Exploration (COMPLEX). These are *A Strategy for the Exploration of the Outer Planets 1986-1996*, *Strategy for the Exploration of Primitive Solar-System Bodies—Asteroids, Comets, and Meteoroids: 1980-1990*, and *Strategy for Exploration of the Inner Planets: 1977-1987*, published in 1986, 1980, and 1978, respectively. NASA's own Solar System Exploration Committee (SSEC), formed in 1980, was charged with providing an implementation plan for the COMPLEX strategies and did so in two reports: *Planetary Exploration Through Year 2000—A Core Program* (1983) and *Planetary Exploration Through Year 2000—An Augmented Program* (1986). The present plan may be regarded as a follow-on, and includes the SSEC plan in its earlier parts.

The task group's recommendations fall naturally into two categories:

- *Completion of the reconnaissance and exploration of the solar system.* By 1995 all the planets except Pluto will have been visited, but many of them for one brief flyby; encounters with asteroids and comets will be just beginning. Many phenomena will remain to be identified, let alone studied and explored.
- *Intensive exploration of Mars.* Mars is the planet most like Earth, and among the most accessible. A sequence of missions, possibly leading to a temporary or permanent base, is recommended.

Technological initiatives that can enable, or greatly improve, activities in various areas are discussed later in this chapter.

Scientific Investigations

The detailed objectives in Table 2.1 can be achieved by the following types of investigation.

TABLE 2.1 Recommended Programs Beyond 1995

Scientific Objectives	Implementation
1. Mercury	
a. geological mapping, gravity, motion, surface chemistry	orbiter, transponder
b. surface and internal properties	lander, sensor network
2. Venus	
a. heat flow, internal properties	sensor network
b. detailed characterization	sample return
3. Moon	
a. detailed study	sample return, sensor network, rover
4. Mars	
a. internal and atmospheric properties	sensor network
b. detailed characterization	rover, sample return
c. field studies	in situ human investigations
5. Asteroids and comets	
a. exploration	multiple rendezvous
b. detailed characterization	sample return
c. field studies (earth-crossing)	in situ human investigations
6. Jupiter system	
a. deep atmosphere investigation	deep probe
b. magnetosphere mapping	polar orbiter
c. surface and interior of satellites, especially Io	lander, sensor network
7. Saturn system	
a. atmospheric investigation	probe, deep probe
b. satellite investigations	orbiter
c. ring investigation	rendezvous
d. Titan atmosphere	probe
e. Titan surface (ocean?)	lander
8. Uranus system	
a. atmospheric investigation	probe, deep probe
b. satellite and ring investigations	orbiter
9. Neptune system	
a. atmospheric investigation	probe
b. satellite investigations	orbiter
c. Triton investigations	to be determined
10. Pluto system	
a. system exploration	orbiter
11. Other planetary systems	
a. search	telescope on Space Station
b. investigation	advanced telescopes in space

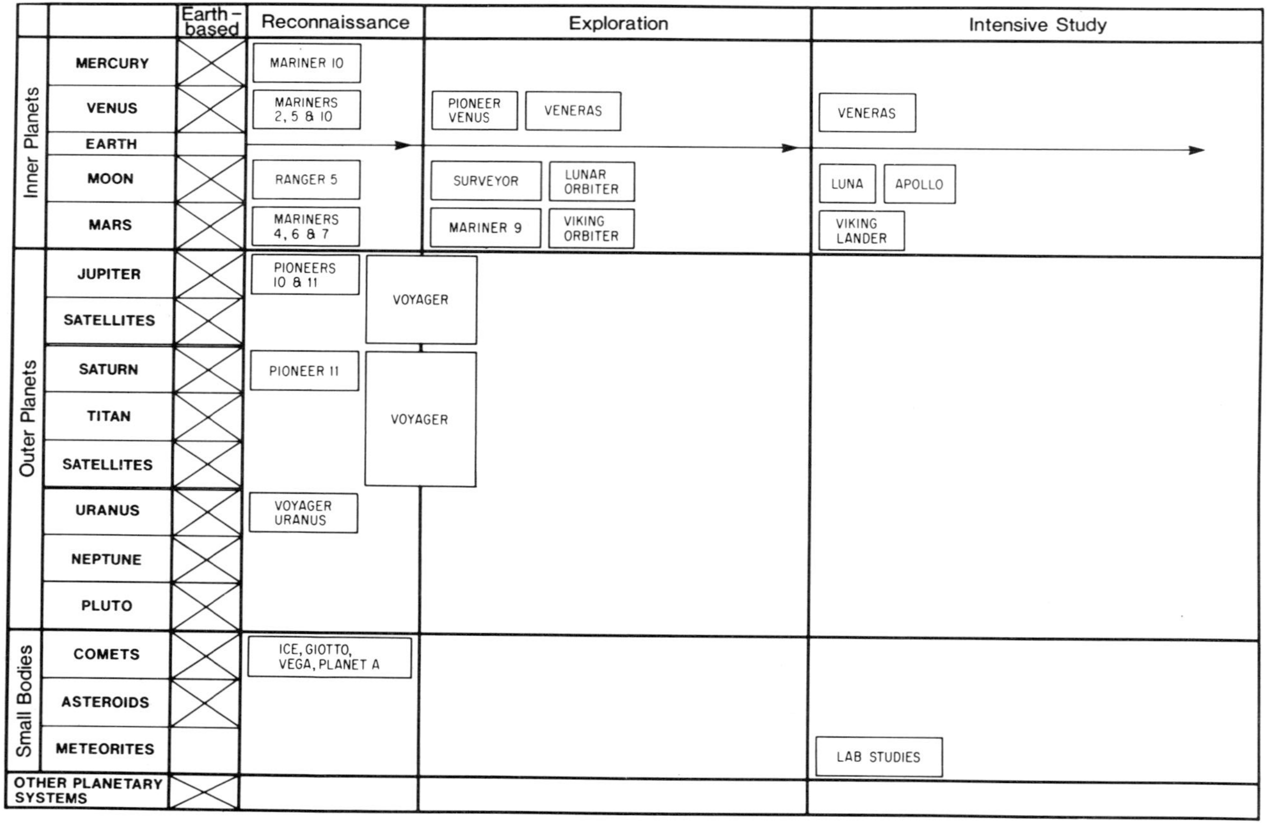

FIGURE 2.1 Planetary exploration to 1986.

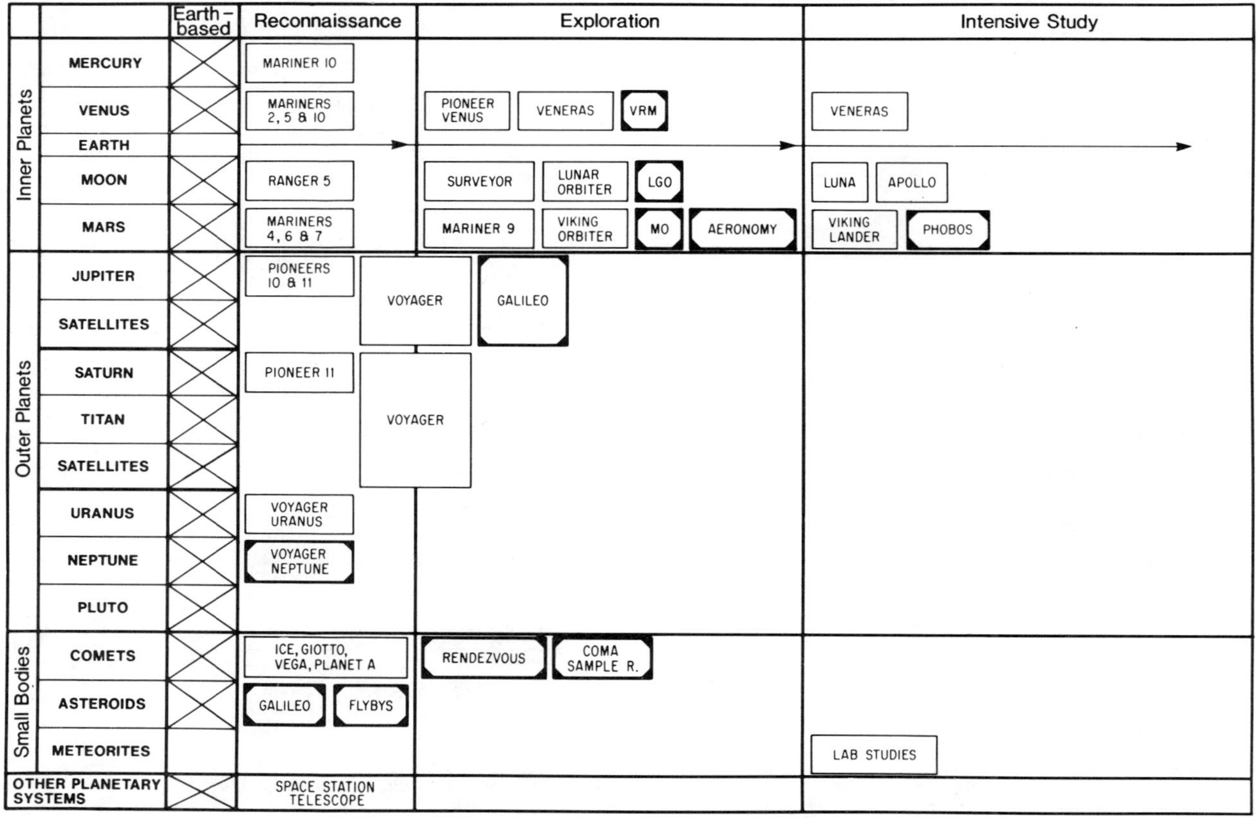

FIGURE 2.2 Planetary exploration to 1995. New missions (since 1986) in boldface.

	Earth-based	Reconnaissance	Exploration	Intensive Study
Inner Planets				
MERCURY	⨯⨯⨯	MARINER 10	ORBITER	
VENUS	⨯⨯⨯	MARINERS 2, 5 & 10	PIONEER VENUS, VENERAS, VRM	VENERAS
EARTH				
MOON	⨯⨯⨯	RANGER 5	SURVEYOR, LUNAR ORBITER, LGO	LUNA, APOLLO
MARS	⨯⨯⨯	MARINERS 4, 6 & 7	MARINER 9, VIKING ORBITER, MO, AERONOMY	VIKING LANDER, PHOBOS, **ROVER, NETWORK, SAMPLE R.**
Outer Planets				
JUPITER	⨯⨯⨯	PIONEERS 10 & 11	VOYAGER, GALILEO	
SATELLITES	⨯⨯⨯		**ORBITER-PROBE**	
SATURN	⨯⨯⨯	PIONEER 11	VOYAGER, **ORBITER-PROBE**	
TITAN	⨯⨯⨯		**ORBITER**	
SATELLITES	⨯⨯⨯		**ORBITER-PROBE**	
URANUS	⨯⨯⨯	VOYAGER URANUS	**ORBITER-PROBE**	
NEPTUNE	⨯⨯⨯	VOYAGER NEPTUNE		
PLUTO	⨯⨯⨯			
Small Bodies				
COMETS	⨯⨯⨯	ICE, GIOTTO, VEGA, PLANET A	RENDEZVOUS, COMA SAMPLE R.	**NUCLEUS SAMPLE R.**
ASTEROIDS	⨯⨯⨯	GALILEO, FLYBYS	**MULTIPLE RENDEZVOUS**	
METEORITES	⨯⨯⨯			LAB STUDIES
OTHER PLANETARY SYSTEMS	⨯⨯⨯	SPACE STATION TELESCOPE		

FIGURE 2.3 Planetary exploration to 2005. New missions (1995 to 2005) in boldface.

15

		Earth-based	Reconnaissance	Exploration	Intensive Study
Inner Planets	MERCURY	✗	MARINER 10	ORBITER	LANDER, SAMPLE R.
	VENUS	✗	MARINERS 2, 5 & 10	PIONEER VENUS, VENERAS, VRM	VENERAS, PROBE, **LANDER NETWORK**, **SAMPLE R.**
	EARTH	✗			
	MOON	✗	RANGER 5	SURVEYOR, LUNAR ORBITER, LGO	LUNA, APOLLO, **ROVER, NETWORK, SAMPLE R.**
	MARS	✗	MARINERS 4, 6 & 7	MARINER 9, VIKING ORBITER, MO, AERONOMY	VIKING LANDER, PHOBOS, **ROVER, NETWORK, SAMPLE R.**
Outer Planets	JUPITER	✗	PIONEERS 10 & 11	VOYAGER, GALILEO, **POLAR ORBITER MAGNETOSPHERE**	DEEP PROBE
	SATELLITES	✗		ORBITER-PROBE	IO LANDER
	SATURN	✗	PIONEER 11	VOYAGER, ORBITER-PROBE	DEEP PROBE
	TITAN	✗		ORBITER	"LANDER"
	SATELLITES	✗		ORBITER-PROBE	
	URANUS	✗	VOYAGER URANUS	ORBITER-PROBE	DEEP PROBE
	NEPTUNE	✗	VOYAGER NEPTUNE	ORBITER-PROBE	
	PLUTO	✗		**ORBITER**	
Small Bodies	COMETS	✗	ICE, GIOTTO, VEGA, PLANET A	RENDEZVOUS, COMA SAMPLE R.	NUCLEUS SAMPLE R.
	ASTEROIDS	✗	GALILEO FLYBYS	MULTIPLE RENDEZVOUS	SAMPLE R.
	METEORITES	✗			LAB STUDIES
OTHER PLANETARY SYSTEMS		✗	SPACE STATION TELESCOPE		

FIGURE 2.4 Planetary exploration to 2015. New missions (2005 to 2015) in boldface.

- *Composition and internal structure* are important for all bodies. Terrestrial planets and large satellites should be sampled at many locations, and networks of seismic and heat-flow stations put in place. Studies of geology and surface geochemistry are important. For bodies with visible surfaces such work can be carried out by imaging from orbiters and landers, and by other remote-sensing techniques, such as infrared and gamma-ray spectroscopy.
- Many crucial types of chemical and isotopic analysis can only be made on *samples returned to Earth*. Such studies bear not only on the present state of crustal material, but also on its origin, age, and history. For Mars and Venus, the samples must be from carefully chosen, well-documented sites. For comets, the main consideration is to preserve the original physical state of the material.
- *Rendezvous missions* are important for the study of small bodies, comets, and asteroids. The behavior of comets, with their changing distance from the Sun, is of special interest. Although such studies may have begun by 1995, they should be continued in order to explore the diversity of comets and asteroids. A rendezvous with Saturn's rings, though technically difficult, would be immensely valuable.
- *Probing of deep atmospheres* is important for all the jovian planets, as well as for Venus and Titan. A start has been made for Venus and Jupiter. However, the Galileo probe to Jupiter will reach only the 12- to 20-bar level—not nearly deep enough to penetrate the main cloud layers of planets such as Uranus and Neptune. Analytical instruments on probes offer the only means of measuring noble gases and isotopic abundances, which are almost the only clues to planetary origin and evolution. Such probing of planets whose surfaces are inaccessible can be regarded as the equivalent of sample return for rocky and icy bodies. In addition, the ability of probes to measure winds and other atmospheric motions is unique.
- *Planetary environments* include magnetospheres, satellites, ring systems, and extended atmospheres. Because of the variety of their interactions, these objects and phenomena are best studied by diverse payloads covering many of the above disciplines on long-lived orbiters, perhaps combined with landers or probes.
- *Other planetary systems* are already being sought, and there are a few tantalizing indications of related phenomena,

such as circumstellar disks. Although there is promise in continued ground-based searches and those made with the Hubble Space Telescope, a dedicated astrometric telescope and a low-light-scattering telescope in orbit will be required for a comprehensive search and follow-up studies.

- *The Mars program* should be aimed at a deeper understanding of the entire planet and its history. After the current generation of orbiters, most further studies will require soft-landing automated laboratories, supplemented by networks of stations for seismic and other studies. Intelligent sample return requires use of rovers, which can carry out geological and perhaps geochemical studies as well. Further geological work requires a human presence, either literally or at the end of a control and communication link, unless a remarkable amount of intelligence can be built into robotic devices.

Technical Developments

Many of the recommended investigations will be enabled or enhanced by technical developments beyond those of the mid-1980s. The power and flexibility of *low-thrust propulsion* make it the key to serious study, beyond the reconnaissance and exploration phases, of comets, asteroids, and the solar system beyond the inner planets. For the jovian planets, the very long trip times of ballistic flights create serious problems, and the capability of getting into orbit is limited.

Closely related is the concern about *power sources*. Though solar power is usable only in the inner solar system, much larger arrays would be useful; large amounts of nuclear power seem available only from reactors, and such systems are already seeing some development.

Quite a different power problem exists at the surface of Venus, where long-lived landers would be important but are currently infeasible because of the 750K temperature. Soft-landing technology, as used on Surveyor, Apollo, and Viking, has been very successful, but is also expensive and terrain-sensitive. Much cheaper systems are needed so that arrays of instruments can be deployed on many bodies; payloads could be relatively modest. Penetrator technology has aroused a great deal of interest for this reason, but other possibilities for *hard or semihard landers* should be explored, as well as the technology for rovers. The planetary program already builds a large degree of autonomy into its spacecraft, but very little ability to make independent decisions. Thus, further development in *robotics or artificial intelligence* would be useful. In addition, some of the most interesting environments in the solar system offer conditions that are intolerable to present-day electronics for more than a few hours. The jovian radiation belts prevent exploration of Io, its torus, and the entire inner magnetosphere of Jupiter. The surface of Venus is very hot. Developments in *radiation-hardened and high-temperature electronics* would improve the chances of exploring these regions.

On-orbit staging, assembly, and fueling offer new capabilities for the more ambitious missions, especially returning samples from Mars. Such missions could benefit greatly from the ability to use a space station to assemble and fuel larger spacecraft than can be launched fully fueled in a single assembly from the Earth's surface.

An integral part of the recommended program is adequate support for analysis and interpretation of the data returned from it, including maintenance of the intellectual base for scientific activity. Theoretical, laboratory, ground-based, and earth-orbital studies must be supported; this support includes upgrading laboratory instrumentation and computing equipment, as well as maintaining adequate programs of research and analysis.

International Collaboration

The matter of international collaboration cuts across the whole of space science, and is therefore considered in the report of the steering group. However, it is appropriate to note here that planetary exploration is a particularly suitable area for such efforts. Planets are a neutral ground devoid of nationalistic, commercial, or military interest. They are analogous to the desert continent of Antarctica, which is protected by an international treaty for scientific research. Modest international participation in national programs has been successful on many occasions, but true cooperative programs, jointly planned by more than one agency, offer great potential benefits.

3
Status of Planetary Science in 1995

OVERVIEW

This chapter begins with two sections summarizing the accomplishments of solar system exploration over the three decades from 1965 to 1995, and the expected scientific questions as of the end of that period. The remaining sections constitute a much more detailed status report for individual classes of objects, with further discussions of open questions. All this material supports, and leads to, the program of missions for the period 1995 to 2015 presented in Chapter 4.

State of Planetary Exploration as of 1995

Among the high points already attained or anticipated for the first three decades of planetary study ending in 1995 are:

- Mercury: Characterization of physiographic provinces for half the surface; discovery of a planetary magnetic field.

- Venus: Establishment of atmospheric and cloud composition; characterization of the high-temperature surface environment; preliminary elemental analysis of surface material from landers; study of solar wind interaction; determination of global topography and gravitational field; characterization of physiographic provinces from radar images.
- Moon: Determination of detailed geological history, chronology, and geochemistry of major geological provinces; detailed study of selected samples of surface material; investigation of cratering, regolith formation, and interaction of the surface with the solar wind for an airless body; discovery of remanent magnetic fields; seismic characterization; measurement of heat flow; determination of composition of the solar wind, both present and ancient. (By 1995, global surface mapping should be achieved or under way.)
- Mars: Near-global mapping of topography, gravity field, and thermal properties; establishment of geological diversity (volcanoes, canyon lands, polar terrains, etc.); discovery of evidence for former extensive surface water (e.g., valley and channel networks); preliminary surface chemical analysis from landers; establishment of structure and chemical and isotopic composition of the atmosphere; determination of geological processes and a relative chronology; study of local and global meteorology over three martian years from landers and orbiters; search for microbial life and organic compounds (yielding negative results). (By 1995, global characterization—morphology, elemental distributions, and some mineralogy—of surface units is expected.)
- Jupiter system: Study of atmospheric composition and circulation; detailed composition and structure of atmosphere and clouds from direct entry probe; discovery of atmospheric lightning and auroras; detailed characterization of the magnetic field and the magnetosphere (sources and sinks, plasma processes); study of the Io plasma torus and of the interactions between this satellite and the magnetosphere; discovery and characterization of the Io volcanoes and interior heat flow; discovery of the ring and several small satellites; comparative studies of icy and rocky planetary objects. (The Galileo orbiter will carry out detailed global mapping of the large Galilean satellites and continue efforts in many of the other areas mentioned above, especially the torus and magnetosphere. The probe will carry out a detailed sounding of Jupiter's atmosphere and clouds.)

- Saturn system: Initial global study of Saturn and its magnetosphere; establishment of atmospheric composition differences between Jupiter and Saturn; detailed study of the ring system and investigation of new dynamical phenomena; discovery of several new satellites, including previously unknown orbital configurations; measurements of the composition and structure of the atmosphere and clouds of Titan; low-resolution mapping of satellite surfaces, except Titan.
- Uranus: Results of Voyager flyby (1986). (Initial discoveries include a strong magnetic field with a large inclination and remarkably diverse geology on several of the satellites.)
- Neptune: Results of Voyager flyby (1989).
- Comets: Results of Halley flybys (1986), including imaging of the nucleus, and exploration of the proximate environment. Also, deployment of planned comet rendezvous missions.
- Asteroids: Results of Galileo flyby of a selected asteroid and of planned flybys by the Comet Rendezvous mission.
- Meteorites: Evidence for early magnetic field, late additions of material with differing nucleosynthetic histories, widespread high-temperature events in the solar nebula; many examples of core formation in small bodies, basaltic volcanism, extraterrestrial synthesis of amino acids; discovery of meteorites from the Moon and possibly Mars.
- Other Planetary Systems: Discovery that many stars are surrounded by dust clouds or disks, and imagery of one such disk; discovery of a star with a planet-like companion. (Many follow-up studies are expected by 1995.)

Applications of these results to the study of planetary origin and evolution include:

- Establishment of the age of the solar system as 4.6 billion years by analysis of radioactive decay products in the Earth, meteorites, and lunar samples.
- Dating of the late stages of accretion of the Moon (and presumably the other terrestrial planets) as 3.7 billion years ago, although most of the mass was probably accumulated within the first 10^7 or 10^8 years.
- Determination of a geological chronology for the Moon, with the final major stages of lunar volcanism measured at 3 billion years ago; establishment of the current rate for impact cratering in the Earth-Moon system.

- Comparative studies of geological processes on the terrestrial planets and the icy satellites of the outer solar system, including impact cratering, volcanic and tectonic activity, and erosional and depositional processes.
- Preliminary study of the development and evolution of planetary crusts in planets of different compositions and internal structures, with insight into the role of tectonics and magmatism in the formation of the crust and interior of the Earth and other planets.
- Inference that the great bulk of the atmospheres of Earth, Mars, and Venus are all secondary, that is, degassed from the interior or acquired late in accretion, and not remnants of the gas from the solar nebula.
- Discovery of unique and as yet unexplained abundances of noble gases (total amounts, relative amounts, and isotopic ratios) on Earth, Mars, and Venus.
- Discovery of a large (100 times) enrichment of deuterium on Venus compared with Earth. Venus must have started out with much more water (or vapor) than it has now, and a "runaway greenhouse" may have caused most of it to be lost.
- Discovery that all terrestrial bodies have experienced differentiation, with accompanying volcanism and tectonics, but with differences in history from one planet to another.
- Discovery of the uniquely high levels of volcanic activity on Io, and preliminary characterization of volcanism based on different physical-chemical systems than had been encountered in the terrestrial planets. In the Saturn system, resurfacing on Enceladus represents yet another example of such volcanic activity.
- Discovery of unexpected complexity in the rings of Saturn and Uranus (e.g., the presence of shepherd satellites, of spiral density waves, and of bending waves), providing important insights into the dynamics of self-gravitating spinning disks.
- In situ investigation of plasma processes of wide astrophysical application in the huge magnetospheres of Jupiter and Saturn.
- The determination of the composition of Jupiter's atmosphere, which is expected to be representative of the composition of the solar nebula, especially for hydrogen and the noble gases. The abundances that will be determined by the instruments on the Galileo probe will probably become the standard for solar composition.

In supporting future investigations, an essential contribution will be made by theorists who endeavor to model the natural evolution of gas-dust disks into stars and their associated planetary bodies. Theoretical investigations of the early stages of this evolution begin with numerical and analytic modeling of star formation, in particular, the conditions under which single stars like the Sun can form. Study of the later stages of this evolution emphasizes modeling the manner and time scale for the accumulation of dust into planetesimals, and the subsequent accumulation of these planetesimals into planetary cores of silicates, metal, and ices. In the case of at least Jupiter and Saturn, the final stage of formation involved the gravitational capture of massive envelopes from the gas of the disk.

Between now and 1995 we can expect that continuing progress will be made in this field, most likely without the help of crucial observations or sudden theoretical breakthroughs. However, in the absence of a new generation of observational facilities that permit higher resolution imaging of other protostellar systems, it is quite possible that in the next decade we will not address the first-order questions required to make substantive progress. On the other hand, we can look forward to a significant refinement and enhancement of theoretical understanding concerning many aspects of nebular evolution. Much of this progress in theoretical understanding is contingent upon the availability of computational resources of continually greater power.

If the first asteroidal flybys occur during the next decade, we can expect to begin to be able to place the great wealth of meteoritical data into a planetological context. We can also expect that basic information regarding early solar system history will continue to flow from laboratory study of meteorites and stratospheric collection of interstellar particles. In this connection, it should be pointed out that, to a large extent, the current laboratory instrumentation used in this work was obtained during lunar sample analysis during the 1960s and early 1970s, and that attention must be given to modernizing the laboratories in which this work is done.

Scientific Questions as of 1995

Fundamental questions in planetary science will remain much the same in 1995 as they are today, but new knowledge and new capabilities will alter our view of how to approach them. First, the reconnaissance and exploration of the solar system will by no means be completed. Saturn and Titan are already ripe for in situ investigation and study of interactions among the magnetosphere, rings, and other satellites. Investigation of comets and asteroids will have begun, but intensive study and exploration of the wide diversity of asteroids will remain. In this area we will want to know the following: the overall structure of the asteroid belt and its radial variations of composition and physical characteristics, which are expected to reveal clues about the structure of the protoplanetary nebula; the mechanisms that powered the evolution of differentiated asteroids; and the chemical composition and physical character of comet nuclei, in order to determine under what conditions these most primitive planetesimals formed.

Internal structure of terrestrial bodies is a broad field for which, apart from the Earth, we still will have only the limited data for the Moon from Apollo, and the even more limited data for Mars from Viking. Even such basic information as crustal thickness will still be lacking. The absolute history of planetary bodies will not be understood without an unambiguous chronology based on radioactive clocks. For example, it is suspected that the martian channels and volcanoes were formed over a protracted period, even though the time scale is based only on crater counts and is very uncertain. There is little prospect of obtaining dates by other means than laboratory analysis of returned samples. Such samples remain valuable long after their acquisition and return to Earth: improved techniques can (and do for the Moon) continue to be applied to the original samples.

Only one side of Mercury will have been imaged from spacecraft, but all the other terrestrial planets are known to be asymmetric in the distribution of geological provinces. While the Galilean satellites of Jupiter will have been studied in some detail, only the most rudimentary reconnaissance will have been made of the other outer planet satellites. Only single flybys of Saturn, Uranus, and Neptune will have taken place, and the Pluto system will remain unvisited.

Our ideas about the origin of this solar system lead us to believe that planet-forming processes occur commonly during star formation. We will want to determine the prevalence and the properties of planetary systems around other stars accurately enough to compare them with one another, as well as with our own system.

We will want to carry on detailed studies of protostars in order to ascertain the physical character of their accretion disks, thought to be the sites of planet formation.

It seems likely that Earth is the only site of organic life in the solar system, but there is no dearth of organic molecules on or in such objects as meteorites, Titan, the jovian planets, and giant molecular clouds located in other parts of the galaxy. Mars, formerly the object of greatest interest, is now seen to be the site of destruction of organic compounds by an intensely oxidizing atmosphere and soil. Conditions, however, may have been more benign in the remote past. There is still much to be learned about the origin of life by study of the objects mentioned above, and perhaps others such as comets. If other planetary systems exist, they may be seats of organic evolution.

PLANETARY GEOSCIENCES

During a relatively short period of time, studies of planets made by earth-based telescopes have advanced to detailed in situ measurements from spacecraft of the planets' surfaces and atmospheres. A complex view of the planets and their satellites continues to emerge.

In late 1962, Mariner 2—the first interplanetary spacecraft—flew by Venus: the journey of Voyager 2 is still in progress. The 203-kg Mariner 2 had only six instruments, whereas the 818-kg Voyager 2 has two color TV cameras and ten other advanced instruments. These two spacecraft represent the simple beginning and the sophisticated continuation of solar system exploration.

In the early years of exploration, missions were selected more by technical feasibility than by scientific priority. So little was known that any mission greatly increased our knowledge. Now, comparative study of the planets is a significant scientific endeavor. Great advances in understanding the origin and evolution of the planets and properties of the solar system will come from comparisons of all planetary objects. Common features such as atmospheres, magnetic fields, and geologic processes can be understood best by such comparison. In turn, these comparative planetary studies provide insight about the history and evolution of the Earth. Nevertheless, exploration has shown that each planet is unique and interesting in its own right.

Scientific Objectives for Planetary Geosciences

The following topics in planetary geosciences contribute to an understanding of the solar system: formation; interior structure, dynamics, and physical state; crustal evolution; and planet morphology and surface processes. These topics, and the measurement objectives for them, are discussed below.

Formation

One key to understanding the formation of the planets is the determination of their chemical and isotopic compositions and the timing of their accretion. The results can be compared for all the planets, satellites, and meteorites in order to place constraints on models of chemical differentiation as a function of heliocentric or planetocentric distance. The results also shed light on the potential for heat sources—important for considerations of internal activity—and to assess models of planetary accretion.

Interior Structure, Dynamics, and Physical State

Measurements of the seismic behavior of planets, the strength and nature of their magnetic and gravity fields, and the heat flow from their interior are critical for determining the characteristics of planetary interiors. When combined with knowledge of mass and composition, the results permit assessment of the nature of possible interior differentiation (core/mantle/crust) and the possibilities for an internal dynamo.

Crustal Evolution

A principal objective in planetary exploration is the determination of the age, composition, and distribution of crustal materials, including volatiles. The results allow refinement of models relating to planetary accretion, differentiation, and degassing. In addition, such determinations allow assessment of the style and timing of volcanism and tectonism and their relation to other geological events, as well as the role of volcanism in the evolution of possible atmospheres.

Planet Morphology and Surface Processes

The types and distributions of landforms and other geological units on planetary surfaces can be determined through geological mapping using remote-sensing data. The results allow assessment of the processes, such as volcanism and tectonism, that have led to the formation and modification of planetary surfaces. Some landforms, such as dunes and valleys, are indicative of processes associated with wind and water, and thus contribute to models of atmospheric evolution. Assessments must therefore be made of the distribution and exchange of volatiles among the crust, regolith, poles, and atmosphere. Knowledge of the geological processes—volcanism, tectonism, impact cratering, and surficial modifications—can be combined with relative and radiometric age determinations of the features associated with those processes to derive geological histories of the planetary surfaces.

An important aspect of comparative planetology relates to the origin and evolution of life. Knowledge of the geological environments permits assessment of the likelihood for the evolution and sustenance of organic life, at least in comparison to Earth. The images of Earth taken from space with its thin skin of oceans and clouds help us begin to appreciate the uniqueness of our planet and the fragile balance that makes life here possible.

An additional impetus for planetary exploration is the potential for using space resources. In a period when natural resources are being depleted rapidly on Earth, no detailed assessment has been made of the resources that exist in space. The Moon and asteroids may hold significant potential as sources of metals and minerals for utilization in space. The initial utilization of such resources may be to support space missions that would travel farther into space, or permanent bases on the Moon or Mars.

Measurement Objectives

The goals outlined above guide the definition of a set of general scientific objectives as follows:

- Characterize the internal structure, dynamics, physical state, and bulk composition of the planet of interest;
- Characterize the planet's chemical composition and mineralogy of surface materials on a regional and global scale;

- Determine the planet's chemical composition, mineralogy, and absolute ages of rocks and soil for the principal geologic provinces;
- Characterize the processes that have produced the landforms of the planet;
- Determine the chemical and isotopic composition, distribution, and transport of volatile compounds that relate to the formation and chemical evolution of the planet's atmosphere, and their incorporation in surface and crustal rocks and polar ice;
- Characterize the planetary magnetic field and its interaction with the upper atmosphere, solar radiation, and the solar wind;
- Determine the extent of organic chemical and possible biological evolution on Mars and Titan, and explain how the history of the planet constrains these evolutionary processes.

The Inner Solar System

The inner planets—Mercury, Venus, Earth and its Moon, and Mars—range from 0.4 to 1.5 AU in distance from the Sun and are smaller and denser than the outer planets. These terrestrial planets are composed chiefly of rock and metal, are poor in volatiles, and have few satellites. Their densities range from 5.4 g/cm^3 for Mercury to 3.9 g/cm^3 for Mars. The variation in density with solar distance has been discussed in the context of a thermodynamic model for the proto-solar nebula in which temperature and pressure decrease with distance from the nebular center and control the chemistry of condensed material. However, it may be that primary differences in planetary density are due to accidental variations in fractionation and reaggregation from collisions. After their formation, all inner planet surfaces were significantly modified by a wide variety of internal and external processes. Nevertheless, each planet has followed its own evolutionary path.

By exploring this diverse family of planets and by comparing their features with those of the Earth, we seek to characterize the evolution of the inner solar system and the causes of the unique aspects of each planet. We also seek to gain insights into the history, as well as the future, of Earth and the life that has evolved on it. Further insights into the terrestrial planets will come from study of the large satellites of the outer solar system as well.

The Moon

Apollo yielded an enormous advance in the understanding of planets by providing samples of the Moon and a wealth of other information. The time scale of the Moon's evolution has been established, and several first-order questions have been answered. Equally important, a basis has been established for interpreting the evolution of other planetary bodies, including the Earth. During the accretionary phase of continuous planetesimal in-fall, the Moon appears to have melted to depths of at least a few hundred kilometers. The ancient crust developed during this maelstrom, with segments repeatedly fragmented and reincorporated into the evolving magmas until a thickness was established that could withstand the waning bombardment.

The larger craters on the Moon record a period of intense bombardment that ended about 3.7 billion years ago, a phenomenon that presumably affected all of the inner planets at about the same time. This bombardment provides a chronological reference, accurately measured in the case of the Moon by radioisotope dating techniques, that is the basis for constructing the geologic history of Mars, Mercury, and (presumably) Venus.

The evolution of the crust of the Moon is known from remote sensing, from instrument data provided by landers, and from study of returned samples. Remote-sensing data show that two major provinces constitute the lunar crust: young (sparsely cratered), low-albedo mare terrains, and old (heavily cratered), high-albedo highlands. The oldest reliably dated rocks on the Moon (from the highlands) are radiometrically dated at about 4.5 billion years old. The youngest mare basaltic lava flows are estimated to be about 2.3 billion years old. The lunar highlands appear to be the result of differentiation at 4.5 billion years and consist of minerals that floated in the melt. However, there is controversy as to whether the upper crust of the Moon was generated in a "magma ocean," whether the whole planet was molten, or whether local areas were successively molten over a long period of time.

Geophysical data show that the Moon had a strong magnetic field early in its history, but the field has since disappeared. Although most scientists consider the Moon to have a small, partly molten core, its presence is the subject of intense debate due to differences in interpretation of the seismic record.

Because all the data provided by Surveyor, Apollo, Soviet

landers, and sample returns are gathered from the Earth-facing hemisphere, there is a need to obtain data from other areas of the Moon in order to establish a better understanding of the geochemical, geophysical, and geologic history. Since the Soviet Union's Luna 24 sample return in 1976, there have been no missions to the Moon. However, a lunar geoscience orbiter (LGO) may fly before 1995. During its one-year mission, LGO would map the global elemental and mineralogical surface composition, measure surface topography, and map the global gravity field. Although no specific date has been set for this mission, the United States, the Soviet Union, Japan, and the European Space Agency are considering LGO-type missions for the 1991 to 1995 period.

Mercury

The geology of Mercury is known primarily from data returned by the flybys of Mariner 10, in 1974. From photogeological studies and remote-sensing data we have determined that the evolution of the crust of Mercury resembles in many ways the crustal evolution of the Moon. Less than half of the surface was imaged by Mariner 10, however, and there may well be other terrains and processes not yet known. Moreover, most of the images are of moderate resolution, on the order of hundreds of meters, and details of the surface are very poorly known.

Most of the known surface of Mercury is heavily cratered and may consist of rocks similar to those of the lunar highlands—differentiated rocks high in silica and alumina. The dominant feature on Mercury is Caloris, an enormous multiringed impact basin. The Caloris basin is partly filled with smooth plains materials interpreted as flood lavas similar to the mare basalts on the Moon. However, the characteristic vents and lava flow fronts seen so clearly on the Moon and Mars are less well displayed on Mercury, possibly due to poor image resolution, and so this interpretation is more controversial.

Apparently, the silicate crust formed early on Mercury—as on the Moon—and heavy bombardment continued. Later flooding of mafic lava flows also occurred, but this may have declined around 3 billion years ago as it did on the Moon. Imaging at higher resolution and coverage for the other half of the planet may discover features indicative of other processes and much younger events. Mapping compositional distributions globally could reveal distinctive provinces and help to explain the high density of the planet.

Venus

Of the inner planets, we know the least about the geological evolution of Venus. Most of the data for this "sister" planet of Earth have come from the U.S. Pioneer-Venus mission, along with Earth-based radar, and Soviet landers (Veneras 8 through 14, Vegas 1 and 2) and orbiters (Veneras 15 and 16).

The surface of Venus is hidden from view except by radar imaging systems and imaging from the surface. Earth-based radar and data from Pioneer-Venus provide an assessment of the gross topography. Venera-15 and -16 radar images have spatial resolution of about 1 to 2 km and cover part of the northern hemisphere, approximately one-quarter of the planet. Two-thirds of Venus is highland terrain, which includes plateaus higher and more extensive than those on Earth. In contrast, the lowland plains ("ocean basins") are only one-third as extensive and one-fifth as deep as the ocean basins on Earth. There are linear mountain belts around Ishtar Terra, the northern "continent," which include the highest mountains, known as Maxwell Montes. These mountains surmount Lakshmi Planum—a plateau that is twice as large and 1000 m higher than the Tibetan Plateau, the largest on Earth.

The Beta and Atla regions, first identified on earth-based radar images, are large shield volcanoes. Beta is composed of high-potassium basalt, while the lowland plains east of it are covered by tholeitic basalts similar to the ocean floor on the Earth and the mare lavas on the Moon. Feldspathic (high silica and alumina) rocks occur in the upland rolling plains that occupy most of the Venus surface. Both Beta and Atla have large gravity anomalies of approximately 135 mgals, corresponding to a compensation depth of more than 100 km, and thus requiring support by an upward flow of mantle material. The indirect discovery of abundant lightning in these areas has been interpreted as indicating the presence of active volcanoes. In general, there is a strong positive correlation between gravity and topography, suggesting that the dominant source is interaction of mantle convection with a surface layer—perhaps a global crust of tens of kilometers thickness. The long-wavelength gravity field of Venus thus contrasts sharply with the

Earth, which is a mix of deep sources (hot spots), plate tectonic, and other effects.

The linear mountain belts may be due to tectonic compression; other belts of fractures may be due to crustal extension. Some circular fracture patterns, called coronae, are more than 600 km across. They may be volcanic-tectonic features, some form of impact-generated structure, or a result of both internal and external processes. Many small craters have been observed, which may originate in impacts or volcanism, although it is difficult to determine their origin with the resolution of present images.

The high deuterium/hydrogen ratio observed by Pioneer-Venus may indicate the loss of substantial water by photodissociation and hydrogen escape. The yellow-gray color of the surface rocks may indicate that the iron minerals in those rocks have been oxidized like those on the surface of Mars, and fine-grained dark deposits may be windblown fine material that partly covers the surface. There are also extensive rift zones that segment the crust. However, these tectonic features do not form an integrated planet-wide system of rifts and ridges, nor are there obvious continent marginal trenches. Apparently, Venus lacks these indicators of integrated plate tectonic motion; thus Venus must get rid of its internally generated heat largely through conduction, aided possibly by "hot spot" volcanism, rift tectonics, and rift volcanism. Venus seems to be closer to the "one plate" planets like the Moon and Mars rather than the multiplate Earth, but in volcanic and tectonic behavior it is more complex than Mars.

The Magellan (Venus Radar Mapper) mission scheduled to begin operation around 1990, will provide near-global images of the surface at 1-km or better resolution. This will enable assessment of the geological processes that have shaped the surface of Venus, and estimates of the ages and sequences of surface units, and internal processes. These images will also allow us to address questions regarding the former existence of liquid water (e.g., ocean shorelines, river channels), possible correlations of topography and gravity, and styles of tectonism and volcanism.

Mars

The martian surface is divided into two roughly equal hemispheres, the southern highlands and the northern lowlands. The southern highlands have elevations from 1 to 10 km above the planetary reference elevation. This region is dominated by large impact craters and several enormous impact basins, of which Hellas and Argyre are the largest. This ancient, heavily cratered terrain may be similar in some respects to the lunar terrae, formed 3.9 to 4.2 billion years ago. However, the martian highlands are partly covered by younger lavas and mantling deposits possibly of sedimentary origin. These highlands also show extensive reworking by wind and water. Although the composition of the highland rocks is not known, they probably include felsic rocks. The northern lowlands consist of young lava flows similar to the more mafic lunar mare lavas, and have been modified by aeolian, fluvial, and periglacial processes.

Two impressive volcanic provinces dominate Mars: the Tharsis region and the Elysium region. The Tharsis region includes Olympus Mons and three other very large (550 km across) volcanoes surmounting the Tharsis "Rise"—an elevated area standing about 10 km high—plus many smaller volcanoes. Olympus Mons, 600 km in diameter and more than 26 km high, is one of the larger volcanoes in the solar system. The Elysium region also includes several large volcanoes, but appears to be older than the Tharsis region. Moreover, the morphologies of the Elysium volcanoes and the lava flows are different from Tharsis and may indicate differences in the style of volcanism or differences in the composition of the magma at the time of eruption.

Although the Tharsis and Elysium regions are impressive, by far the greatest extent of volcanic rocks occur as various flood lavas and lava plains in the northern lowlands and the southern cratered terrain. Still other large volcanic deposits may represent high-silica eruptions and may include ash flow tuffs that have been heavily eroded by the wind.

The polar areas are covered by extensive, thinly layered deposits, probably of sedimentary origin, that are associated with the permanent and ephemeral ice deposits. The perennial ice caps may be water ice in the north and carbon dioxide in the south. In addition, a vast dune field surrounds the north polar region. Dunes also occur in isolated fields elsewhere on the planet, where they partly fill many of the craters.

Many parts of Mars have been extensively modified by faulting. Valles Marineris is a canyon system more than 4000 km long that resulted from rifting and other tectonic processes. Extensive layered deposits are visible in the sides of the canyons and in mesas within the canyon system. Some of these deposits may have arisen in vast lakes that once filled the canyons. Other theories suggest that some of the layered deposits are lava flows, volcanic explosive deposits, or wind-laid sediments. Regardless of origin, the deposits may represent a large part of martian geologic history, just as the deposits in the Grand Canyon of Arizona represent a substantial fraction of Earth's history.

Some regions of Mars are dissected by large and small channels cut into young rocks, indicating that liquid water existed late in martian history. Other more degraded channels dissect ancient terrain and appear to have formed earlier in martian history. Thus, there may have been many periods in Mars' past when liquid water could exist on the surface.

Despite several successful missions to Mars, including the two Viking landers, many fundamental questions regarding its present state and geological history remain unanswered. The Mars Observer mission will return data in the early 1990s, providing global maps of the surface composition, details of the topography, and data on the lower atmosphere. However, questions about the interior (presence of a core, nature of the mantle, etc.) and of active surface processes will remain unanswered.

Internal Characteristics of the Inner Planets

All the inner planets, including the Moon, underwent significant early heating, melting, and differentiation, but the evolution of the Moon and Mercury terminated early as heat was lost rapidly due to their small size. Like Earth, Mars and Venus are presumably sufficiently large that they are losing internal heat slowly. Their heat sources have continued to operate over billions of years, manifested at their surface in the form of volcanic and tectonic features.

Earth's surface continues to evolve dynamically. Crustal material is continually created at mid-ocean ridges and destroyed beneath deep-sea trenches, as the plates that make up the Earth's crust move in more or less steady relative motion. The formation of mountain belts, the development of volcanic chains, and the driving force behind many large earthquakes are linked to these plate motions. Neither the Moon, Mercury, nor Mars shows global tectonics of such vigor; the surfaces of the Moon and Mercury are old and preserve a record of early heavy cratering and ancient volcanism. The surface of Mars also shows heavy cratering modified by wind and water erosion, but also demonstrates an extensive history of volcanism and tectonism. Data for Venus, although limited, show that this planet also underwent differentiation and has experienced tectonism and volcanism.

The collective study of the inner planets implies that all have been melted and internally differentiated, leading to core, mantle, and lithosphere. Earth's interior is known from seismic measurements to be layered, a product of global differentiation. At Earth's center is a metallic core, largely fluid and in convective motion, but with a small solid inner core. The core is surrounded by a mantle of ferro-magnesian silicates, mostly solid and in very slow convective motion. At the surface, the mantle is capped by a thin, rigid lithosphere of mostly igneous and metamorphic rocks, overlain by a veneer of volcanic rocks and sedimentary material. Each of the other terrestrial planets is thought to be similarly layered, but the evidence is limited.

Fundamental unresolved issues in inner planet studies are: (1) the nature of the convective motions that drive plate tectonics on Earth and the importance of such tectonic processes early in Earth's history, (2) the character of tectonism on other planets and satellites, and, (3) the causes of the major differences in evolutionary style.

Magnetic Fields of the Inner Planets

Earth has a substantial dipolar magnetic field of internal origin, evidently produced by the action of a hydro-magnetic dynamo sustained by motions in the fluid core of the rotating planet. Of the other inner planets, only Mercury has a magnetosphere comparable in character with that of Earth, though much smaller in size. The Moon shows evidence for an early complex magnetic history, now recorded in the remnant magnetism of lunar rocks and the lunar crust, but the origin of the field is not known. Slowly rotating Venus apparently has no internal magnetic field. The existence of a martian magnetic field is debatable because of scanty measurements; however, if a field exists, it is small. The wide differences in the nature of planetary magnetic fields are not understood but may be related to rotation rates and the nature of the core.

Characterizing the particle and field environment, including internal magnetic fields, is important for understanding the solar wind's interaction with a planet. Earth's field extends through a volume of space many times larger than the planetary volume, forming an umbrella that shields Earth from the interplanetary plasma; in contrast, the solar wind blasts the Moon directly. Although Venus does not have an internal field, there is a complex interaction with the solar wind that may be similar to that of comets.

Atmospheric-Climatic Connections of the Inner Planets

The geological record and measurements of planetary atmospheres provide clues to the evolution of surfaces and climate history. The Viking entry and lander measurements of nitrogen isotopes in the martian atmosphere indicate that large quantities of volatiles have been lost from Mars. It is inferred from this and other measurements that Mars has outgassed and subsequently lost to space or to permafrost cold traps the global equivalent of a depth of some tens of meters of water. Because of the planet's low atmospheric pressure, liquid water cannot exist on Mars at present. However, there is strong geologic evidence for the earlier presence of liquid water on Mars and the suggestion of a hydrologic cycle and thus a long, perhaps episodic history of free water on the martian surface.

NASA's Mars Observer, planned to gather data in the early 1990s, will shed light on key questions regarding the present and past interactions between the atmosphere and lithosphere. This mission is proposed to provide: (1) a global view of the distribution of elements and minerals on the martian surface, (2) the water vapor, clouds, and dust distribution in the atmosphere and the vertical temperature profile, and (3) the distribution of ice on or near the surface. In addition, the Mars Observer will obtain global radar altimetry data that are crucial to the integration of the image, gravity, thermal, and chemical information.

A Mars aeronomy mission, possibly flying at the same time as the Mars Observer, could explore the planet's upper atmosphere and interaction with the solar wind and answer long-standing questions about Mars' internal and external magnetic fields. In addition, such a mission could provide data on the net mass exchange between the atmosphere and the solar wind, and provide important clues regarding the history of Mars' present and past atmospheres by determination of the rates at which various volatile elements escape from the martian ionosphere.

Rocky Satellites

The Voyager results made possible the geological study of a host of new objects, ranging from relatively large, silicate bodies to small satellites composed predominantly of ice. Collectively, these objects show surfaces that have experienced impact, volcanic, and tectonic processes similar to the inner planets. In addition, they show processes not seen, or at least not fully appreciated, on the inner planets, including volcanism induced by tidal heating, sulfur-driven volcanic eruptions, deformation of surface features through slow flow of ice-rich crusts, and resurfacing through the eruptions of ice-rich materials.

Voyager results also caused a reassessment of any notion that internal planetary activity is simply correlated with planetary size. Larger objects contain more radioactive constituents relative to their surface area, and hence generate more heat to drive planetary tectonic activity. Thus, it was thought that small bodies would cool and freeze quickly and have short, simple histories of internal activity in comparison to large bodies. Although this seemed to be the case with the inner planets, the concept was drastically modified by Voyager observations. Io, a body the size of Earth's Moon, had nine volcanic eruptions in progress during the encounter, making it the most internally active body in the solar system. The heat to drive these volcanoes is likely derived predominantly from tidal stresses created by Jupiter and the nearby satellite Europa, rather than from radioactive decay, as on larger planetary bodies. Some small satellites of Saturn (notably Enceladus) and of Uranus (such as Miranda) show evidence of resurfacing and extensive tectonism. This is indicative of internal activity. On the other hand, some other larger and smaller satellites appear to have less vigorous or no internal activity.

Many of the questions raised by Voyagers 1 and 2 during their brief encounters with the Jupiter system will be addressed by the Galileo spacecraft scheduled to arrive at its destination in the mid-1990s. During its 20-month mission, Galileo will obtain better estimates of the chemical composition and physical state of the satellites, along with data on magnetic field and particle fluxes in the Jupiter system. The cratering record, the nature of volcanic processes on Io and possibly Europa, and the styles of resurfacing and tectonic processes on Europa and Ganymede will all be substantially better known after Galileo.

Studies of the outer planet icy satellites with their solid crusts and possibly mobile mantles represent an opportunity to address the fundamental problems of the physics, chemistry, and geology of deformed crusts. They will also allow us to study the internal constitution of bodies that differ radically from the inner planets. By 1995, data will be available for a range of bodies, from those with thoroughly deformed crusts to those that are minimally disturbed. With additional geophysical and geochemical data we may be able to verify the causes of the formation of the core, mantle, and lithospheres of these bodies. We may also be able to characterize the lithosphere, asthenosphere, and mantle, and their interactions to produce tectonism and volcanism in terrestrial bodies. Subtle differences in chemistry and phase are necessary to make the terrestrial systems function. Study of systems that involve water-ice crusts and liquid-water mantles may broaden our appreciation of these fundamental problems.

Satellites of Jupiter

The most important advance in understanding satellite evolution arose from the observations of the Jupiter system by Voyager. This understanding will be further advanced by the Galileo mission. The Galilean satellites of Jupiter form a well-ordered family of bodies that cover a wide range of internal dynamism. The inner two, Io and Europa, are about the size of Earth's Moon. Their densities, 3.5 and 3.2 g/cm^3 respectively, indicate that they are probably composed primarily of silicate materials. The outer two, Ganymede and Callisto, are about the size of Mercury and appear to be mixtures of silicates and water, indicated by their densities of 2.0 and 1.8 g/cm^3, respectively.

As discussed above, the discovery of active volcanoes on Io by Voyager makes it the most volcanically dynamic body known in the solar system. Internal heat appears to be generated by the tidal stresses in Io, and the eruptions, apparently driven by sulfur dioxide, reach heights of 250 km. In addition to the products of active volcanic explosions, the surface of Io is dominated by other volcanic features, including a variety of domes, calderas, collapsed depressions, and digitate lava flows that radiate from volcanic centers. The absence of impact craters at the resolution of Voyager images indicates that the surface of Io is very young. Evidently, resurfacing by lava flows and deposits from volcanic emissions is taking place very rapidly.

The composition of the surface material on Io is enigmatic. The presence of sulfur is indicated by spectral reflectance data and various "hot spots" having temperatures consistent with molten sulfur. However, pure sulfur has insufficient strength to form large, steep landforms. The presence of mountains as high as 8 km and scarps up to 1.5 km indicates the probable presence of silicate rocks.

The surface of Europa has very high albedo. This probably indicates an ice or ice-rock crust, which is often densely fractured. Europa is also thought to have a water substrate and a rocky interior. Only a few impact craters have been identified, and, like Io, the surface is considered to be very young.

Ganymede displays two fundamental surface units: an older, heavily cratered, dark terrain and a younger, brighter unit that has been extensively modified by fracturing and other tectonic processes. Impact craters exhibit both bright and dark ejecta; many of the craters have been deformed by viscous flow of the icy crust moving under its own weight. However, some craters retain their topographic expression, and it has been proposed that there was higher heat flow early in the history of Ganymede, which allowed viscous creep to occur at a high rate. Later cooling led to a more rigid crust.

The outermost Galilean satellite is Callisto. Its surface is very heavily cratered and includes several multiringed structures. Callisto does not appear to have experienced any tectonic deformation or volcanism.

Satellites of Saturn

The satellites of Saturn are more diverse and irregular in their crustal evolution than those of Jupiter. The most interesting is Titan. It has low density (1.9 g/cm^3) and a thick atmosphere (1.5 bars) of predominantly nitrogen, plus methane and minor constituents. Clouds masked views of the surface from Voyager; consequently, little is known about the crustal evolution of Titan.

Enceladus is a particularly intriguing satellite of Saturn. It displays ancient, heavily cratered terrain, a crust that is broken by faults, and areas that have been resurfaced. Its density (1.2 g/cm^3) is consistent with water ice, but substantial amounts of methane and other ices may also be present. Enceladus, like Io, may have been volcanically active as a result of tidal stresses, giving it a crust and mantle that may still be active.

The other Saturnian satellites are heavily cratered and are so cold and rigid that the craters retain their original topographic form and are not viscously deformed, as are the craters on Ganymede and Callisto in the Jupiter system. They exhibit different degrees of internal activity, varying from rift valleys, as on Tethys, to faulted and partially resurfaced crusts, as on Dione. Bright wispy zones, some of which are associated with faults, may reflect water frosts erupted from fissures.

Iapetus has a larger variation in albedo (dark to bright) than any other satellite in the solar system. It may be affected by dark material swept up from orbit, emplaced by impact, or deposited via internal activity. The current data are of highly inadequate resolution to resolve this question.

Satellites of Uranus and Neptune

The Voyager 2 encounters of the Uranus system in early 1986 returned the first images and other data on the nature of the satellites. Ground-based studies have already demonstrated that the five known satellites of Uranus are icy objects, perhaps similar in bulk composition to the satellites of Saturn, but with darker (dirtier) surfaces. All five satellites were imaged by Voyager 2 at resolutions of a few kilometers or better, but the mission emphasis was on the innermost known satellite, Miranda, which was encountered at close range, yielding subkilometer resolutions, as good as any obtained by Voyager of the satellites of Jupiter and Saturn.

The 1989 Voyager encounter with Neptune is being planned with that planet's largest satellite, Triton, as a prime target. The spacecraft will fly close to Triton and will provide an occultation as seen from Earth. The occultation is particularly important since Triton is known to have an atmosphere.

Ground-based telescopic studies have also revealed evidence of liquid nitrogen and perhaps hydrocarbons on the surface, as well as frozen methane, making this nearly lunar-sized object potentially one of the most interesting members of the satellite family.

ATMOSPHERES

Atmospheres of planets and moons exist in an enormous variety. The terrestrial objects Venus, Mars, and Titan have atmospheres somewhat similar to Earth's, even though they span several orders of magnitude in surface pressure. The jovian planets, including Jupiter, Saturn, Uranus, and Neptune, have extremely deep atmospheres of which we can expect to explore only the outermost skin. They are dominated by hydrogen and helium rather than the oxygen, nitrogen, and carbon dioxide of the terrestrial planets. Very tenuous atmospheres are found on Mercury, Io, comets, and probably a few of the icy satellites of the jovian planets; in many ways they resemble the outermost parts of the denser atmospheres, exhibiting phenomena such as escape and the presence of ionization.

Atmospheres are studied to determine their present state—their composition, structure, meteorology—and also to find clues to their origin and evolution. To first order, the gases in the jovian planets resemble those in the Sun, and are therefore taken to be of primary origin, with little change since their formation. Such gases are almost, but not quite, absent on the terrestrial planets; instead we find "volatiles" that could plausibly have accreted as components of solids. Later degassing and chemical alteration would then produce what is found today. For example, photolysis of ammonia and loss of hydrogen to space give nitrogen. Likewise, life on the Earth has converted carbon dioxide to oxygen, organic molecules, and buried carbon in the form of carbonate rocks. The traces of "primary" gases, neon, and heavier noble gases can only be measured from within the atmosphere or on a sample. In this area our present information is limited to Earth, Mars, Venus, and the parent bodies of certain meteorites. The most remarkable thing about these results is their diversity, which has so far resisted any attempts at an overall explanation. Nevertheless, it is important to know that primary atmospheres either never existed on bodies as large as the Earth or were almost entirely lost.

Earth, Mars, and Venus

The atmospheres of Earth, Mars, and Venus have all evolved markedly from their initial states. The other planets show that biological activity, which seems to have dominated on Earth, is not the only agent that can have a profound effect on a planetary atmosphere or surface.

Viking measurements of the Mars atmosphere showed that a large fraction of the original nitrogen has been lost. The abundance of water must have been enough at one time to cut large numbers of fluvial channels; current surface temperatures and pressures do not allow water in the liquid state. Mars may have lost an amount of water equivalent to a layer 10 m deep to space and to sinks below the surface, although some resides in the polar caps.

Even larger amounts are missing from Venus, as shown by the huge enrichment of deuterium in the atmosphere measured by Pioneer Venus. Depending on the exact style of loss, as much as an earth ocean of water could have been lost over the life of the planet. This could explain the immense underabundance of water on Venus relative to the Earth, since the planets are otherwise quite similar. However, the original amount is very poorly determined, and the site of the oxygen left behind by the escaping hydrogen has not been established. It is usual to assume that the loss took place early in Venus' history. However, this timing should be determined and any possible relation to a change in the tectonic style of the planet should be explored.

The history of Earth is similar to the histories of other terrestrial planets. Common aspects include early global differentiation of crust and core, outgassing and evolution of an atmosphere, and early bombardment of the surface by a heavy flux of meteoroids. Of course, Earth has many attributes not shared by any other planet. These include its oceans, its high oxygen abundance, its tectonic motions and the consequent complex history of crustal deformation, its life forms, and its development of a global magnetic field and magnetosphere.

Earth is the only planet that has large quantities of free water on its surface and in its atmosphere. The dynamics of Earth's oceans play a large, incompletely understood role in the regulation of the terrestrial climate. Earth is also unique in the large quantities of molecular oxygen in its atmosphere as the result of biological activity. Venus makes a startling contrast: it is covered by a dense global blanket of clouds composed of sulfuric acid droplets and has a thick, hot atmosphere of carbon dioxide. Cloud motions and tracking of descent probes and balloons indicate a global wind pattern with substantial dependence on height for the mean wind speed. Surface winds are mild, but 100 m/s winds blow at the cloud tops. Martian winds are variable, as on Earth, with annual episodes of high-velocity winds that often give rise to global dust storms. Mars also has marked seasons, with carbon dioxide cycling between the polar caps driving a major component of atmospheric circulation. Layered sedimentary deposits at the martian poles are evidence of long-term climatic changes, whose origins are poorly understood. On Earth, such climatic changes have given rise to the periodic ice ages. Earth is the only planet whose surface, atmosphere, and hydrosphere have provided an environment conducive to the development of life and the evolution of complex living organisms. These life forms have substantially influenced the chemistry of the atmosphere and the oceans and the major sedimentary rocks on Earth's surface.

The Viking mission showed the absence of detectable organic molecules on Mars. It also revealed that an intense ultraviolet flux from the Sun reaches the surface. This suggests that living organisms are not present on Mars now. Whether Mars was less hostile to the development of life during earlier times, when it may have had a denser atmosphere and flowing surface water, is still an open question. Other basic questions about Mars are the fate of all the missing water and the nature of the current hydrological cycle linking the polar caps, ground water and ice, and the atmosphere.

A rare group of meteorites, called SNC for the initials of places where the first examples were found, are widely believed to be samples from a martian lava flow. Most remarkably, they contain gases whose elemental and isotopic composition is substantially identical to that measured by the Viking landers, within the errors of the latter.

Study of the present state of an atmosphere is carried out both in situ and remotely by flybys and orbiters. Except for Mars, all the terrestrial bodies with substantial atmospheres exhibit a greenhouse effect: the surface temperatures are higher than they would be in the absence of an atmosphere. There is a great deal of interest in this phenomenon from the standpoint of past and future climates on the Earth, where we are in the midst of a grand experiment of the effect of massive injections of carbon dioxide from human activity. Clouds, smogs, and dust are interesting both in their own right and as tracers of atmospheric motions. Observed clouds include water and ice on Earth, ice and solid carbon dioxide on Mars, ammonia, methane, and perhaps liquid water or ice on the jovian planets. Smogs (produced photochemically rather than by condensation) are found on Earth, Venus, Titan, and at least some of the jovian planets. Dust is very important on Mars and is significant on the Earth as well.

Theoretical meteorology of the Earth is finely tuned to our observational knowledge. It has been extended to Mars with considerable success, but is very unsatisfactory with Venus and Jupiter. On Venus we observe winds near the cloud tops (about 65-km altitude) blowing from the east at 100 m/s, and similar winds are thought to exist in the region of the ionosphere. Not only were these motions unpredicted, they still cannot be explained in any fundamental way now that they have been observed. On Jupiter and Saturn, similar velocities are seen, but with large shears between zones. These are just as far from being explained.

Progress in meteorological theory will rely on a combination of further theoretical and computational work, and more and different observations. A theory with true predictive power would be especially important in the area of climatic change. Improved short-term forecasting, however, is more likely to be obtained by still-more-detailed observations of the Earth itself.

The study of photochemistry and photoionization is much more readily transferable between planets. Mars and Venus are excellent Earth analogs. The dayside ionosphere of Venus is well understood, and the stratospheres exhibit similar phenomena. Limited data on the Mars ionosphere suggest that it is similar to that of Venus, but measurements from a long-lived orbiter are needed. Catalytic ozone destruction is much more important in the stratospheres of both Mars and Venus than on Earth. Both Venus and Earth have sulfate layers, but again this layer is much denser on Venus. Further study of these planets should continue to shed light on terrestrial pollution problems.

The nightside upper atmosphere and ionosphere of Venus are unexpectedly cold. The corresponding region of Mars is unexplored and may or may not be similar. Comparative study should cast light on this mysterious phenomenon.

Titan

Titan, the largest satellite of Saturn, is unique among satellites in having a dense atmosphere: predominantly nitrogen, but also containing a small amount of methane and possibly argon. The surface pressure is 1.5 bars. Since the surface gravity is only 135 cm/s^2, the amount of gas per unit area is nearly 11 times that of the Earth.

A nearly uniform orange haze hides the surface and any condensation clouds that might exist. This orange haze is likely composed of condensed hydrocarbons, nitrile compounds, polyacetylenes, and HCN. The precipitation of these compounds may form a deep layer lying on the surface, or they may be dissolved in a liquid ocean. Dense methane clouds probably lie some distance above the surface. This rich inventory of organics in a nitrogen medium provides a natural laboratory for the study of prebiotic organic chemistry, which may be relevant to the question of the origins of life on Earth.

The origin of nitrogen on Titan is a fundamental, unsolved question. One theory is that the nitrogen could be primordial, incorporated in a clathrate ice during accretion. Two other sources have also been suggested: photolysis of ammonia and high-temperature formation from ammonia during impacts. The other atmospheric constituents can be derived by photochemistry from a nitrogen-methane mixture, except for the traces of carbon monoxide and carbon dioxide. The carbon monoxide could be outgassing from the interior, or could be derived from the ice in incoming meteoric material, with the carbon coming from the atmospheric methane.

Titan is embedded in a torus of escaped gases, which includes atomic hydrogen, observed by Voyagers 1 and 2, and probably molecular hydrogen and nitrogen. Ionized torus material contributes to the plasma in Saturn's magnetosphere: impact by magnetospheric particles is an important loss process for the neutral torus. Voyager 1 passed through Titan's magnetospheric wake and observed a number of changes in the plasma and magnetic environment.

The existence and nature of the predicted ocean of ethane and nitrogen (or alternatives, such as lakes or puddles) should be studied, both remotely and in situ. There are important interactions of such an ocean with the atmosphere, for which it would be a source of volatiles and a sink of photochemical products. Of the volatiles, methane is a likely candidate to produce the dense cloud layers, whose presence is strongly suspected, but which have not been confirmed because they are hidden by the photochemical smog.

An atmospheric circulation, especially strong in the stratosphere, has been inferred from the temperature field observed by Voyager 1. Testing by direct wind measurements could tell us whether our understanding, based on Earth and Venus data, is good enough to encompass Titan as well. Finally, the hydrogen torus surrounding Titan's orbit has provoked speculation about the presence of other atoms and molecules in the torus, and the torus' interactions with the magnetospheric plasma. Also of interest is the possibility of sources for the torus other than Titan itself; Saturn is a possible candidate.

Io and the Plasma Torus

Io, Jupiter's innermost large satellite, exhibits such remarkable phenomena as volcanoes believed to be driven by sulfur dioxide, a tenuous atmosphere of sulfur dioxide, a persistent extended cloud of sodium atoms, and a plasma torus containing ions of sulfur and oxygen enveloping the orbit. In turn, many of these ions become energized and populate the magnetosphere, and probably drive the large escape rates that populate the torus itself. The energy source for the vulcanism is tidal heating by Jupiter and nearby satellites. Most of the other aspects are poorly understood at best, and some are totally mysterious. For example, although it seems clear that sulfur dioxide is passing through the atmosphere to the torus, there is no agreement on the density of the atmosphere.

Study of all these phenomena would be near the top of any priority list if it were not for the unbearable environment of the enveloping magnetosphere. Pioneer and Voyager instruments making a single pass have been damaged, and radiation damage is a primary concern for the Galileo spacecraft and its instruments. A prolonged stay near Io does not seem possible with current or easily foreseen technology.

The most burning question about Io's atmosphere is the actual quantity of sulfur dioxide and its variations with latitude and time of day. The amounts of other gases, such as oxygen, are also of great interest. A large pressure bulge is expected near the subsolar point, and some models predict strong winds blowing away from this region. Probably the most remarkable thing about this atmosphere is the huge escape fluxes of sulfur, oxygen, and sulfur dioxide that populate the torus, and that require replacement in a very short time. The mechanism itself is obscure, and there is the possibility of large variations with time. Among the atoms involved is sodium, which gives a bright glow easily observed from Earth. It must be a substantial component of the crust, but the chemical form is unknown.

Although the ionic composition of the torus is fairly well known, the energy sources are not fully established. Since this object drives much of the rest of the magnetosphere, it needs to be understood as well as possible. Another energetic phenomenon related to Io is the intense radio (decametric) bursts, which have been studied from Earth for decades but whose origin is still being discussed.

Jovian Planets

From their very low mean densities we know that Jupiter, Saturn, Uranus, and Neptune have extremely deep atmospheres, extending perhaps more than halfway to the centers. They also possess cores of 10 to 20 earth masses similar to large terrestrial planets, but possibly much richer in the ice-forming compounds water, methane, and ammonia. A large percentage of the atmospheres appears to be made up of these compounds; there is strong evidence for methane and ammonia, but much weaker indications of water. They also form cloud layers, perhaps more than one per planet, but the highest layer (ammonia on Jupiter and Saturn, methane on Uranus and Neptune) tends to obscure the deeper ones. Particularly for Jupiter, the cloud patterns are arranged in bands parallel to the equator, instead of the cyclonic whirls found on Earth. Several simple organic molecules and a dark stratospheric smog, also seen on Titan, are believed to be produced photochemically from methane.

Jupiter, Saturn, and Neptune all have internal heat sources that are very large by terrestrial standards, and comparable to the heat they receive from the Sun. Uranus' source may be almost as large as Neptune's, but it cannot be resolved from the reradiated solar heat. Such a flux from the interior is expected to have profound effects on the structure and circulation of an atmosphere, but the details are not understood. The heat source itself is of major interest; it is probably a combination of residual heat from planetary formation and gravitational energy from continued differentiation and rainout of heavier elements, such as helium, due to immiscibility with hydrogen. The former heat source dominates on Jupiter and the latter on Saturn. Precise in situ measurements of the helium-to-hydrogen ratio in the atmospheres of Saturn, Uranus, and Neptune will contribute to an understanding of these processes.

The dynamical properties of the interior are closely coupled to the composition, structure, heat flow, and rotation of the planet. Convective flow of conducting matter generates an external magnetic field that reflects some of the properties of the internal flows, with changes in the flow producing secular changes in the magnetic field. These properties of planetary magnetic fields can be best determined from orbiting spacecraft with a small periapsis.

Differences in the bulk composition of the outer planets are related to differences in the temperature, pressure, and chemistry occurring at different radial locations in the solar nebula at the time of planetary formation and to the nature of the accretion and collapse process. Because the major species such as carbon, nitrogen, and oxygen are present mainly as methane, ammonia, and water, which are deficient in the upper atmosphere due to condensation into clouds, their abundances must be determined by probes below the cloud decks. On Uranus and Neptune, it is thought that the water cloud base will lie below 100 bars.

It is often presumed that the different-colored clouds of Jupiter are at different altitudes and indicate regions of upwelling and downwelling. However, there is little direct evidence that this is the case and there are other indications from observations of atmospheric scattering properties that are difficult to reconcile with such a model. Synoptic multiband observation with moderate spatial resolution would be valuable, as would in situ observations of cloud depths in different latitudinal regions.

Internal structure is related to the origin and evolution of the planets, since it depends on bulk composition, on the accretion process, and on subsequent evolution. It is also of interest for the insight it provides into the properties of matter at high temperatures and pressures. Knowledge of the bulk composition will be essential, and improved values for the higher gravitational moments, as can be derived from an orbiting spacecraft, would be useful. Laboratory and computer studies of high-pressure properties are also essential.

It is possible that the observed zonal wind flows are an atmospheric skin effect extending only as deep as sunlight penetrates (several bars). Alternatively, the zonal winds may reflect an internal flow pattern extending deep into the neutral atmosphere, driven by the internal heat source. In situ wind measurements with a simple atmospheric probe, and synoptic optical, infrared, and microwave observations from an orbiting spacecraft, may provide relevant information.

RINGS

Planetary rings, once thought to be unique to Saturn, have been observed around all the giant planets except Neptune. Even for Neptune, there is evidence for the existence of partial rings. The ring of Jupiter is optically thin and composed of dustlike small particles. Saturn's rings are broad, bright, and opaque, whereas the rings of Uranus are narrow and dark. They all lie predominantly within the Roche limit, where tidal forces would destroy a self-gravitating body, and also within the planetary magnetosphere.

The goals of the study of planetary rings include three major objectives. The first is to understand their composition, active processes, and origin. A second objective is to study their active processes as analogs of those that operate in other flattened, rotating, dynamic systems like galaxies, accretion disks, and our own solar system at an earlier stage. A third objective is to study the particles as remnants of an earlier stage of solar system evolution; they are not as primitive as the comets, but less processed than the larger planets and satellites in the outer solar system.

The best-studied ring system is Saturn's, which has been observed from the ground for centuries. The most detailed information on planetary rings is from Voyager spacecraft observations. Ground-based radar, photometry, and infrared spectroscopy have been complemented by spacecraft imaging, spectroscopy, and occultation. We now have a reasonably comprehensive inventory of the ring material surrounding Jupiter, Saturn, and Uranus, and a preliminary understanding of some important dynamic processes in each of these systems. Continuing theoretical modeling using existing data sets will focus questions about the physical processes that govern the morphology and stability of planetary rings.

The occurrence of rings around the massive planets is evidence of an evolutionary path parallel to planetary aggregation. The nonlinear, dynamical interactions responsible for forming and flattening planetary rings also operate in galaxies, planetary systems, and stellar accretion disks. Generally, rings consist of planetary material that accreted originally with the planet. This material was either never incorporated into larger bodies, or may have formed satellite bodies that were later broken up. In both Saturn's E-ring and the jovian ring, there is Voyager evidence for short particle lifetimes, arguing for continuous replenishment from satellite surfaces. A fundamental open question is the age of the Saturn ring system: the proximity of the inner satellites is inconsistent with the action throughout the age of the solar system of current dynamic processes, which push them away from the rings.

Saturn's rings are mostly water ice and emit thermal radiation in energy balance with incident sunlight. Radar, radio occultation, and spectrophotometric studies indicate that particles larger than 1 cm are responsible for most of the Saturn ring opacity. The composition and thermal state of Jupiter's and Uranus' rings are unknown, but interaction with magnetospheric ions at Jupiter may both heat and differentiate ring material. Remote sensing has a limited capability to characterize ring particle composition and size: by 1995, any spectroscopically active constituents should have been identified in both the solid and gas phases. In the subsequent period, a ring rendezvous mission could provide in situ analysis of particle and ring-atmosphere composition.

Ring particles display complex collective interactions. Voyager Saturn data show mass clustering into thousands of ring features. Density and bending waves, spokes, and even multiple strands have also been observed. Nine distinct, narrow rings have been identified at Uranus. Many of the Saturn structures are gravitationally induced by satellite orbital resonance; for example, the outer ring edges of Saturn's brighter ring occur at radial distances where particle orbital periods are commensurate with those of satellites. Small, close satellites "shepherd" the F-ring. A satisfactory explanation does not exist for the multitude of smaller-scale Saturn ring features, and explanations of the broad structure of Uranus' rings and Jupiter's ring are unconfirmed. Since ring particles may acquire net charge by either ultraviolet or charged particle irradiation, electrodynamic forces influence the motions of the smallest particles, probably producing the Saturn ring spokes and likely limiting the lifetime of the small particles in Jupiter's ring. A ring rendezvous could measure the plasma environment of Saturn's rings in the vicinity of a spoke feature.

The orbit pole of an inclined elliptical ring precesses in response to the higher moments of the planetary mass distribution. This effect has been observed for Uranus' epsilon ring, and has provided a value of J_2, the second gravitational moment, with a small uncertainty. By 1995, improved values for the even the multipole moments of giant planet gravity should be available from improved ring and satellite astrometry.

In the mid-1990s, we will have new observations of the jovian ring from the Galileo orbiter and observations of Saturn's rings (including stellar occultations), from both earth-based and earth-orbiting instruments. We will also have data from Voyager studies of the uranian rings, and will have determined at least upper limits on any Neptune ring material. In addition, we expect that a Saturn orbiter mission will be en route to Saturn to arrive at the turn of the century. For the jovian and saturnian systems we can hope to have some long-term, detailed information on dynamics and secular changes. We also expect to know about the spatial variation of important observables like composition within those ring systems. However, the details of individual particles and their mutual interactions still may be hidden, and we will still lack any in situ measurement of their composition and local environment. The origin and nature of both the large-scale and the small-scale structure in Saturn's rings are likely to be resolved only when the composition, physical properties, and size distributions of particles in the A-, B-, C-, and D-rings are adequately characterized.

Among the dynamical processes believed to be important in Saturn's rings are direct collisions, gravitational scattering, diffusion and angular momentum flow, resonance interactions with various satellites and Saturn, electrodynamic processes, and possibly diffusional instabilities. The exact nature of these processes is not understood, however. For example, although the outer edges of the A- and B-rings are known to result from satellite resonances, the edges are much more abrupt than expected. A more complete understanding of the nature of the processes and their importance depends upon knowledge of inelastic collisions between ring particles and of the particles' relative velocities. There is currently no direct information on either of these.

Although it is generally agreed that the outer edge of the A-ring is maintained by the outward transfer of angular momentum to the small co-orbital satellites, the resulting dynamical time scale for the expansion of the orbit of these satellites is short, suggesting either that they are transferring angular momentum to another satellite through some unrecognized mechanisms or that the outward flow of angular momentum from the A-ring is smaller than thought. The dynamical time scale for the inner edges of the rings also poses a problem, since there is no known mechanism for transferring angular momentum to the ring particles to halt their inward spiral toward Saturn.

The thickness of the rings is dynamically determined by the balance between energy lost and the random energy supplied during collisions. This clearly depends upon the relative importance of the various interparticle interactions, including direct collisions and gravitational scattering. These two physical processes are distinct and result in different dependences of the vertical distribution of particles on their relative velocities, size distribution, and physical properties. The vertical spread of the largest particles could be quite different from that of the smaller particles.

The overall composition of the ring material is related to the origin of the ring, while any segregation of the material according to particle size or location involves dynamical processes that have either produced or maintained the segregation. Closely related to the composition is the physical nature of the individual particles. They may be fluffy or solid, cohesive or merely short-lived aggregations of smaller pieces. Observations of individual particles from a ring rendezvous could provide this information.

Models for the formation of the observed radial spokes involve unobserved aspects of the plasma and neutral particle environment of the rings. Although synoptic studies from a Saturn orbiter will considerably improve our knowledge of the kinematics of spoke creation and dissipation, measurement of the ring environment by a ring rendezvous would directly address the nature of the processes involved.

For the uranian rings, answering current questions will involve providing an inventory of the ring material and associated satellites, determining the relative role of self-gravity and collisions in creating the morphology of the elliptical rings, and discovering waves and other small-scale structures not visible from earth-based stellar occultations. All of these were addressed by the Voyager Uranus encounter in 1986. This, however, will be the last look at the uranian rings before 1995.

It is not possible to predict what will be the most important questions arising after the new Voyager results, since it is likely that we will find that some of our current concepts are incorrect. However, a number of major questions remain open after the Voyager Uranus flyby. These are the detailed dynamics, long-term changes, composition, and spatial variation of the rings. Further, we will need to know their neutral and charged-particle environment. One discovery is that the fringes of the atmosphere actually envelop the rings and cause a very substantial drag—yet another indication of a short time scale. Clearly, a long-lived Uranus orbiter is needed to address these questions.

Jupiter's ring will be extensively imaged by the Galileo orbiter before 1995. The Galileo observations will help answer important questions for this system. The CCD (charge-coupled device) camera on Galileo is more sensitive than Voyager's and will make multiple observations of the jovian ring and small satellites nearby. This will refine our knowledge of the morphology of the ring and better define the particle size distribution. We expect increased understanding of the evolution of individual particles under the combined effects of plasma and gravity. We hope to clarify the relation between the ring particles and the small moons embedded in it, which appear to be supplying ring material.

Since the life history and lifetime of the small jovian ring particles are strongly affected by the plasma environment, deeper understanding requires directly measuring that environment. This requires in situ measurement of the inner jovian magnetosphere, most plausibly using a Jupiter orbiter with a low perijove.

INTERIORS OF THE GIANT PLANETS

Our knowledge of the giant planet interiors is indirect and relies heavily on theoretical plausibility arguments. The procedure involves the computation of theoretical models with several adjustable parameters that are varied to match observed quantities. These quantities include the mass, radius (and hence the mean density), and often several of the higher gravitational moments that depend both on the rotation period and the internal distribution of the mass.

Prior to spacecraft flybys of many of the giant planets, the radii of Jupiter and Saturn were known only from remote measurements, the masses were fairly accurately known from planetary perturbations, and the first two gravitational moments were approximately known from satellite perturbation; the observational situation for Uranus and Neptune was much worse. The flybys that have taken place through the Jupiter, Saturn, and Uranus systems have much improved the accuracy of all these quantities, and meanwhile there has been an extensive effort to determine these quantities better for Neptune by remote observations. By 1990 we should also have spacecraft flyby observations for Neptune.

The construction of interior models depends upon having a good equation of state, which, in turn, depends upon a knowledge of the elemental constitution of the interior. It has been customary to assume that the major elements, hydrogen and helium, should be present in solar proportions, since there is no known way in which the planets can be formed that would result in the chemical fractionation of these two gases. Models composed of hydrogen and helium alone are deficient, and do not fit the observations; the discrepancies indicate that there is an excess of material of higher mean molecular weight near the center. The mass excess indicated is approximately 10 earth masses for each of the giant planets. This is a large excess relative to the planetary mass for Uranus and Neptune, and so can be obtained from even crude knowledge of the planetary parameters, but it is a much smaller fraction of the planetary mass for Jupiter and Saturn, so that the higher precision of the knowledge of the planetary parameters has been essential in those two cases. We can say very little about the distribution of this excess mass, other than that it is core-concentrated, and we cannot determine directly whether it is entirely rocky in composition or whether there may be a major contribution to it from "ices" such as water or ammonia.

One major theoretical question is the solubility of helium in hydrogen in the interior. Remote sensing measurements from the Saturn flybys have indicated a reduced helium/hydrogen ratio in the Saturn atmosphere compared to Jupiter. This has been interpreted as indicating insolubility of helium in hydrogen on Saturn, where the interior temperature is lower than in Jupiter. Thus it is expected that helium has "rained" out of the Saturn envelope to lower depths. A consequence of this is an additional gravitational energy release on Saturn, which may account for its observed infrared luminosity, much higher than would be expected from a straightforward cooling history since the planet was formed and contracted to its present size. By the mid-1990s, we expect to have direct helium/hydrogen abundance measurements for the Jupiter atmosphere from the Galileo entry probe, and can be looking forward to similar measurements for Saturn, which will improve the accuracy of this important measurement.

A major question of the interior physics of Jupiter and Saturn is the nature of the transition interface from molecular hydrogen to metallic hydrogen, which is expected to occur at a pressure in excess of one megabar. All of the giant planet interiors are expected to be convectively unstable, which would normally keep the envelopes well mixed. However, this interface to metallic hydrogen may locally suppress convection, and this may help to maintain a disparity in the helium/hydrogen ratio in the upper and lower envelope of Saturn. Related to this may be the puzzle that an internal infrared luminosity has not been detected for Uranus, but one is readily found for the very similar planet Neptune. These are among the important questions that can be addressed through improved modeling, which, in turn, depends upon better measurements of envelope composition and gravitational moments.

Another theoretical expectation is that the cores of the giant planets, whether of rock or ice composition, would dissolve into the overlying hydrogen at the central pressures of the planets, if given an opportunity to do so (unless the rocky material arrives in very large chunks). The implication is that they have not been given a chance to do so, and hence the mass excesses in the cores have not been produced by infall of material into the assembled planets and settling through their envelopes. This points toward a process of formation of the giant planets in which the cores formed first and then captured the hydrogen and helium. This, in turn, places a number of important constraints on the history of the solar system, and it requires confirmation, not only through the construction of better interior models, but also through better laboratory measurements of the relevant physics.

In the mid-1990s, we expect to know the relative abundances of hydrogen, helium, carbon, nitrogen, and hopefully oxygen in the jovian atmosphere as a result of the probe entry in the Galileo mission. It should be possible to make better measurements of the higher gravitational moments because of the Galileo mission. The heat flow from the interior of the planet and the configuration of the jovian magnetic field should be measured somewhat better.

These will allow improved models of the planet to be constructed, but our knowledge of the planetary core will remain very crude.

For Jupiter it will be desirable to have a follow-up mission that includes as a component a spacecraft in polar orbit with at least part of the orbit lying close to the planet. This will make possible higher-precision measurements of the gravitational moments and will allow a complete mapping of the magnetic field, including the determination of many of the higher magnetic moments. If the abundance of oxygen is inadequately determined in the Galileo mission, then a follow-up probe designed to obtain this highly important parameter will also be desirable. Laboratory studies of the equations of state of the relevant materials and of the solubility of heavier elements in hydrogen at higher pressures will contribute to an understanding of the structure of the interior. Both laboratory and theoretical studies of the metallic-molecular transition in hydrogen will be critical to the interpretation of the jovian interior. A repeat of the polar orbital measurements after an interval of several decades will be important for studying changes in the higher magnetic moments of Jupiter, and hence possibly deducing some aspects of the internal motions of the gas in the envelope at the depth at which the jovian magnetic dynamo operates.

All of these measurements are also desirable at Saturn. A planetary probe is essential to measure the helium/hydrogen ratio accurately and to obtain the abundances of carbon, nitrogen, and oxygen relative to hydrogen. Polar orbiters are needed to measure gravitational and magnetic moments of the planet and to measure the time variations of the latter. The interpretation of the interior structure will be assisted by the laboratory measurements mentioned above.

For Uranus and Neptune only a preliminary flyby of the planets will have been accomplished by 1995; this occurred quite far from the planetary surface in the case of Uranus, but will be much closer in the case of Neptune. That will give us an improved understanding of the structure of these planets. However, hydrogen and helium make up only a relatively small part of the mass of these two planets, and so it is still highly desirable to obtain the best possible measurements of the higher gravitational moments to obtain the interior mass distribution, and to measure the envelope composition, the heat flow from the interior, and the magnetic moments and their time variations with considerable accuracy. This will require entry probes and orbiting spacecraft for these planets beyond 1995.

PLANETARY MAGNETISM

Although the existence of Earth's magnetic field has been known for hundreds of years, it is only during the twentieth century that magnetic fields have come to be known as a phenomenon widespread throughout the universe. In the early part of this century, sunspots were discovered to contain intense magnetic fields. During the past several decades magnetic fields have been found associated with the rest of the Sun, the interplanetary and interstellar media, many of the planets, galaxies, and a large number of stars. There is indication that the Moon was strongly magnetized early in its history and that the protosolar nebula from which the planets formed also possessed a strong magnetic field. Indeed, it is now clear that the absence of a magnetic field in a large cosmical object is the exception rather than the commonplace.

Of the planets investigated so far, only Venus and Mars have no, or very weak, magnetic fields. Mercury, Earth, Jupiter, Saturn, and Uranus all have strong fields. Most of these fields are predominantly dipolar, approximately centered within the planets, and nearly aligned with the planets' rotation axes. The most startling exception is Uranus, having a dipole highly inclined with respect to its rotation, and displaced far from the planet's center. Although such a large departure from symmetry is unusual among known planetary fields, astronomical studies indicate that many stars have magnetic fields with marked departures from symmetry. The surface intensities of known planetary magnetic fields typically fall in the range of a few tenths of a gauss to 10 G at the planetary surfaces, although Mercury's magnetic field is weak in comparison to those of most of the other planets—a few thousandths of a gauss.

A distorted magnetic field stores energy and exerts forces on the medium in which it is embedded. Thus strong magnetic fields can influence the dynamical behavior and evolution of cosmical systems. The rapid release of energy stored in magnetic fields causes explosive, flaring outbursts in many systems, ranging from planetary magnetospheres to the solar corona and high-energy astrophysical objects.

The persistence and behavior of cosmical magnetic fields are now understood to result from similar physical phenomena occurring in a variety of different objects. The general characteristics of objects possessing magnetic fields are that they are large, electrically conducting, and rotating fluid bodies. The ability of an object to retain a magnetic field depends on a combination of large physical scale and high electrical conductivity. Most cosmical objects are large by their very nature, and they conduct electricity in extensive metallic or gaseous ionized regions. The generation of a magnetic field results from the organizing influence of rotation on the convective motions of an electrical conductor. Magnetic fields are ubiquitous in the universe because most natural objects have these attributes, and because large quantities of free magnetic charge, capable of rapidly short-circuiting magnetic fields, are absent.

In addition to being important targets of scientific scrutiny in their own right, planetary magnetic fields provide significant clues to the interior states of planets. In view of the fact that magnetic fields seem to be very easily generated when the requirements described above are satisfied, the absence of a magnetic field in a planet imposes a severe constraint on the state and the motions of its interior fluid. Moreover, detailed observations of a planetary magnetic field can yield important information about the characteristics of the generating fluid motions. Indeed, it was detailed studies of the temporal behavior of the geomagnetic field that provoked early ideas about the motion of fluids in Earth's core and that provided the earliest insights into the regeneration of magnetic fields in natural objects through the phenomenon of the hydromagnetic dynamo. More recently, the discovery of a magnetic field in Mercury was one of the unexpected discoveries that resulted from detailed spacecraft investigations of that planet. The fact that Mercury possesses a magnetic field poses still unresolved questions about its interior structure and thermal evolution.

Generation of Planetary Magnetic Fields

Our present theoretical understanding of hydromagnetic dynamos gives us a strong foundation for understanding the generation and dynamical behavior of magnetic fields in the wide variety of cosmical objects that possess them. However, a great deal remains to be learned. Most of our current understanding is based on linear kinematical theories, which are strictly confined to those situations in which the magnetic field does not itself strongly influence the state of the system. But fundamental considerations and observations indicate that many natural magnetic fields do not operate in such a simple regime. Rather, many systems develop to a fully nonlinear state in which the magnetic forces are as large as any of the others. Advances in our understanding of magnetic field generation and behavior rely on a combination of theoretical and observational studies. Because of the extreme nonlinearity in their behavior, a variety of states may be available to hydromagnetic systems even with similar geometries and boundary conditions. In this respect, the behavior of hydromagnetic dynamos is analogous to that of most complicated cosmic systems.

In a hydromagnetic dynamo, the magnetic field is produced by large-scale electrical currents flowing in a fluid conductor. The electrical currents are, in turn, generated by the fluid's motion across the magnetic field lines. A self-sustaining hydromagnetic dynamo is said to occur when the fluid motion is sufficiently vigorous that new electrical current is generated at a rate that compensates or overcomes the dissipation of the current by electrical resistance. The ultimate energy source for dynamo magnetic fields derives from the forces that drive the fluid convection.

This magnetic field generation process is essentially similar to the mechanism whereby electricity is produced in electrical generators. However, because the fluid systems that occur in natural bodies have so many degrees of freedom, the complexity of behavior of natural dynamos and the variety of states in which they can exist are vast in comparison with the simplicity of electrical generators.

Strictly speaking, the fluid motions in dynamos cannot generate magnetic fields from scratch. The dynamo process is one of field maintenance through regeneration and amplification. In order to get the process started, an initial magnetic field is required to be present, although it may be arbitrarily weak. Such weak magnetic fields are omnipresent, generated by thermoelectric currents and by a variety of other astrophysical and cosmological phenomena.

Not all fluid flows are capable of maintaining magnetic fields. However, a large variety of flows are capable, and most natural objects satisfy the criteria for producing such flows. As already noted, regenerative fluid motions occur in objects that rotate and convect. Most cosmical objects of planetary size or larger satisfy both criteria. The motions need not be very vigorous if the fluid is highly electrically conducting. Earth's magnetic field is generated in the planet's liquid iron core, which has an electrical conductivity not atypical of cosmical objects. The regenerative fluid motions in Earth's core are estimated to be less than a millimeter per second, corresponding approximately to the velocity amplitude that theoretical calculations show to be necessary.

Because of the essential organizing influence of rotation on dynamo fluid motion in rotating objects, and because of the important role played by differential rotation, the close correspondence between dynamo magnetic fields and planetary rotation poles is not mysterious. Because large-scale planetary magnetic fields result from correlated regenerative action at smaller scales, the fact that most fields are nearly centered in their respective planets is also not hard to understand. Moreover, deviations of planetary magnetic fields from the "ideal" centered and axially aligned structure can be understood on the basis of random variations in the fluid motions, although other effects may also occur. In fact, the geomagnetic field—the only planetary magnetic field for which detailed, long-term measurements are available—deviates in a random fashion from the centered, axial ideal, with the time scales of variability being some thousands of years. Averaged over very long times, the geomagnetic field is highly centered and axially aligned with the pole of rotation.

Uranus' highly tilted and eccentric magnetic field challenges the simplest theoretical models of field generation. However, the significance of this challenge is difficult to assess without measurements of the field's possible variation with time.

Mathematical analysis of the magnetohydrodynamic dynamo process shows that magnetic field states of several types can occur. The fields can be generated with a variety of spatial structures and time variations, depending on the geometrical structure of the fluid motion and its amplitude. Some dynamo modes are stationary with time, while others grow or decay monotonically; another class of dynamo modes consists of fields that oscillate periodically and migrate through the generation regions. The polarity of a cosmical magnetic field is not dictated by the dynamo process; if a field can be maintained by dynamo action, then the same field, but with opposite polarity, can be maintained equally well.

Magnetic fields that are observed in natural objects clearly correspond to these calculated behaviors. The Sun's magnetic field seems to correspond to an oscillating and migrating dynamo mode. Earth's magnetic field seems to correspond to a low-order stationary solution of the dynamo equations. For the other planets, too little is known about the temporal behavior to guess whether any of their magnetic fields are oscillating. The oscillation periods are probably too long to observe directly. Uranus' magnetic field is most provocative in this respect. The correct interpretation of its unusual spatial structure depends strongly on whether the field is in a stationary state, an oscillating state, or a transient configuration deviating markedly from a more regular average structure.

Earth's Magnetic Field

For obvious reasons, Earth's magnetic field is the best studied of all cosmical magnetic fields. Along with the magnetic field of the Sun, it constitutes the most important paradigm of cosmical magnetic field generation. Detailed global measurements of the contemporary geomagnetic field have been made for over 150 years—a period of time adequate to discern the dominant features of the field's secular variations. A substantial record of the geomagnetic field's long-term behavior has been extracted from paleomagnetic studies of magnetized rocks and from the magnetization patterns preserved in the sea floor as new oceanic crust cools and spreads away from the mid-oceanic ridges. This paleomagnetic record yields a fairly complete history extending several hundred million years into the past, although, because of sampling problems, the record is probably not reliably complete to time scales less than several tens of thousands of years.

Perhaps the most startling behavior recorded in the geomagnetic record is the phenomenon of geomagnetic reversal. From time to time, the geomagnetic field spontaneously and suddenly changes its polarity. Intervals between polarity reversals are of random duration—averaging some 200,000 years, but with a nearly Poisson distribution. In the intervals between reversals the field exists in a quasi-static, but noisy, state. The actual reversal events proceed very quickly, taking only some 5,000 years. This time is far shorter than the free decay time of the dipole mode, thus requiring that reversals involve an active dynamical mechanism rather than a passive shutting down and restarting of dynamo regeneration within Earth.

The geomagnetic field also exhibits occasional large, but apparently short-lived, excursions from its normal structure. Because of the difficulty of extracting from the paleomagnetic record detailed structural information about short-lived events, these transients are not well mapped. However, it is possible that the geomagnetic field might occasionally resemble what we have recently measured at Uranus.

Studying Planetary Magnetic Fields

It is probably safe to say that the fundamental basis of magnetic field generation in planets is understood. The physical and mathematical machinery of Maxwell's equations and Newton's laws has led to a dynamo theory of magnetic field generation that accounts for basic aspects of the existence and behavior of known magnetic fields. However, beyond that, our understanding quickly fades to a frontier of unanswered questions.

A hydromagnetic dynamo—consisting of a coupled system of fluid and magnetic field—has many degrees of freedom, leading to a highly complex set of possible behaviors. It is not feasible to explore theoretically the full range of mathematically possible behaviors. An effective approach to understanding requires both theoretical study and detailed observations of actually occurring magnetic fields. Theoretical investigations create our understanding of the possible varieties of dynamo magnetic fields, their behaviors, and the conditions under which these occur. Detailed measurements of magnetic fields guide our knowledge about which among these theoretical possibilities are actually realized in physical systems and provoke a more penetrating theoretical understanding by revealing unanticipated phenomena.

Because magnetic fields are generated by fluid flows, a major deficit in our understanding of any natural magnetic field results from continuing uncertainty about the detailed physical character of the fluid flows themselves. In the specific case of planets, ignorance of the state of interior fluid and its motion inhibits the predictive power of even the incomplete theories now in hand. The sources of energy and buoyancy that drive fluid convection in planetary interiors are not well understood. Even for Earth, it is not yet known what is the motive force that drives convection in the core fluid; the most popular candidates are radioactive heat generation and differentiation of heavier materials toward the center.

Most of our present theoretical understanding of magnetic field generation is confined to the so-called kinematical theory—or, equivalently, the weak-magnetic-field limit of the theory—in which the fluid motion is negligibly perturbed by the magnetic field's Lorentz force. It is possible that many peculiar behaviors of natural magnetic fields will only be understood on the basis of more complete theories that can take into account the effects of the magnetic forces. So far, the mathematical tools with which to accomplish this are in a rudimentary state. This problem is likely to be overcome only by application of the largest computers now coming into use.

Observationally, it is important to extend our knowledge of the natural magnetic fields existing in the solar system and the variety of their behaviors. Up to now we have been surprised at every turn, for the most part by the startlingly pervasive occurrence of cosmical magnetism and its astounding variety of forms. Important advances in understanding will come from the combination of information obtained about the magnetic fields of planets with the much greater detail of knowledge that can be extracted from Earth and the Sun. What understanding has been achieved for the Earth and the Sun has largely resulted from our ability to make measurements of the temporal variations, as well as the spatial structures, of those magnetic fields. Because of its accessibility, Earth is a crucial target for detailed study of its magnetic field behavior; what is learned from studying Earth will influence our understanding of all cosmical magnetic fields. It is especially important to ascertain the spatial structures and time dependence of the geomagnetic field on both large and intermediate scales, and over both short and geological times. Measurements of the temporal variations of other planetary magnetic fields are also of great importance. Because of the commonality of physical processes involved in cosmical magnetic field generation, deep understanding will be achieved only from an integrated approach.

It should be noted that Jupiter, Saturn, and Uranus have large and dynamic magnetospheres. Exploration of these objects should go hand in hand with intensive study of the Earth's magnetosphere and tail. This area is discussed in the report of the Task Group on Solar and Space Physics.

PRIMITIVE BODIES AND THE ORIGIN OF THE SOLAR SYSTEM

A major goal of space science is understanding how the Sun and planets were formed. Progress toward achieving this goal involves synthesis of knowledge obtained from many sources, including missions to planetary bodies, astronomical observations from earth-based and earth-orbiting observatories, and geological, geophysical, and geochemical studies of the Earth itself. Knowledge from these sources is then combined with theoretical investigations that link observational data to fundamental physical and chemical laws and processes.

A special role is played by observations of bodies that have been relatively unaltered since the formation of the solar system: comets, asteroids, and the meteoroidal and meteoritic fragments derived from these bodies. Subsequent sections in this report address our anticipated state of knowledge of these bodies in 1995, the principal questions that are likely to be unanswered at that time regarding these bodies, and the programs and technical needs required to address these questions.

It is not appropriate to discuss the broader questions of origin in this report. However, the task group has provided a discussion of how work on this large problem may be expected to proceed. There emerges from this discussion a proposal that does involve specific technical requirements related to the planned space station. This is the search for planetary systems other than our own. This exciting opportunity represents a unique advance in our approach to planetary studies, and is fundamental to future progress in understanding how our own planetary system and life on Earth began.

The Origin of the Sun and Planets

How the solar system was formed is an unresolved question fundamental to planetary science. Answering this question is a prime objective of the NASA planetary program. Serious attempts to devise a theory for the origin of the solar system go back more than three centuries.

During the last decade a qualitative change has developed in our approach to this problem. This has largely been stimulated by the great wealth of information returned from lunar and planetary missions, by astronomical observations of young stellar objects, and from laboratory investigations of extraterrestrial material: meteorites, lunar samples, and interplanetary dust particles. What once was a field of science populated by a few lonely thinkers defending complete but idiosyncratic theories of solar system origin is now a field characterized by a healthy interplay among theory, observation, and experiment.

In association with this change, there has been significant evolution in thinking concerning the origin of the solar system. The picture that is emerging is one in which formation of disks of dust and gas is an essential component of the star formation process. Whether stars form out of such disks, or whether the stars are the principal source of the disks is not yet clear. It is now generally accepted that these disks provide the birthsite for planetary companions to stars. While this general picture is similar in a general way to previous "nebular" models, there is now a growing body of observational evidence that supports it.

Another principal characteristic of current work on solar system origin is its strongly interdisciplinary nature. The formation of circumstellar disks is inferred from infrared and radio astronomical observations of cool, dense interstellar clouds, and density irregularities (called "cores") contained therein. Aspects of stellar astrophysics are also relevant because meteoritic studies show that the early solar system contained significant quantities of now-extinct, short-lived radioactive isotopes produced in stellar nucleosynthesis, such as ^{26}Al, that were rapidly mixed into the proto-solar nebula, and may have been an important heat source in the earliest solar system. In addition, other products of stellar nucleosynthesis were incompletely mixed into the material from which the Sun and planets were formed, as evidenced by other "isotopic anomalies" in meteorites.

Knowledge obtained from lunar and planetary missions is also central to building our understanding of earliest solar system history. The detailed chemical and isotopic compositions of the various planetary bodies—including satellites, comets, and asteroids—are particularly relevant because of their bearing on our understanding of processes and thermodynamics within the solar nebula. Sometimes these planetary observations match in a natural way current thinking of early solar system history. An example of this is the evidence for extensive early cratering found on the less geologically active planetary bodies. This record fits with that expected from the sweep-up of residual planetesimals following the principal stage of formation of these planets. However, unsettling surprises also occur. Perhaps the best example of this was the discovery by Pioneer and Venera spacecraft that Venus' atmosphere contains 50 to 100 times as much ^{20}Ne and ^{36}Ar as Earth's atmosphere. Not only was this unpredicted by all present theories of planet formation, but even now reconciliation of these observations with theory is in an unsatisfactory state.

Meteorites represent another highly rewarding source of data relevant to solar system formation. Most meteorites are fragments of rock and metal, broken off during collsions among asteroids. Detailed laboratory chemical, petrographic, and isotopic investigations of this material continue to provide a wealth of information regarding conditions that prevailed in the primordial solar nebula, as well as of processes that occurred during the formation and subsequent history of these small planets.

Finally, theorists endeavor to model the natural evolution of gas-dust disks into stars and their associated planetary bodies. At present we are far from achieving anything that might be called a definitive understanding of this evolution of a gas-dust disk into a planetary system. On the other hand, on a number of aspects of the problem there are developing "shared understandings" that are likely to evolve into a consensus. An example is the general agreement that the Sun and planets evolved from a disk of dust and gas on a time scale of $\sim 10^8$ years. It is also generally recognized that achieving an understanding of the mechanisms by which angular momentum, energy, and mass are transported during the evolution of this disk is essential. A number of investigators are engaged in theoretical studies directed toward evaluating alternative physical processes (e.g., turbulent viscosity or gravitational torques) by which this transport is effected.

At the opposite end of the sequence of events leading from a gas-dust disk to planets is the final accumulation of the Earth and other terrestrial planets from planetesimals. A considerable quantity of analytic and numerical work has been carried out on the dynamics of this stage of planetary formation. Although many unanswered questions remain, the answers to several important questions appear to be quite model-independent (i.e., not dependent upon whether the accumulation took place in the presence of nebular gas or after the dissipation of this gas). If the terrestrial planets did indeed form from the accumulation of planetesimals, it is found that these planetesimals must have included quite large objects, greater than 100 km in diameter, and possibly as large as the Moon or even Mars. Accumulation of such large bodies onto a planet will result in high initial internal temperatures, inasmuch as the mechanisms for radiating into space the large quantities of energy resulting from the impacts of bodies this large are inefficient. Geologically important corollaries are the prediction of extensive partial melting during the formation of the Earth, primordial chemical differentiation, and formation of the Earth's iron core simultaneously with the formation of the Earth.

With regard to the formation of the outer planets, serious fundamental problems exist. It seems most likely that solid cores with masses of about 15 times the mass of Earth formed first, which then accumulated gas from the nebula. There are very good, if not compelling, reasons for believing that these cores formed before the final accumulation of the terrestrial planets. However, at least in their most simple form, present theories of planetary accumulation lead to the opposite conclusion, that the accumulation of the outer planets required much more time than was required for the growth of the terrestrial planets. Promising suggestions that could resolve this paradox have been made and are now awaiting serious theoretical evaluation.

If the first asteroidal flybys occur in the next decade, we can expect them to begin to help us place the great wealth of meteoritical data into a planetological context. This should make a significant contribution to our ability to evaluate such questions as the extent to which the available meteoritic sample is representative of the asteroidal region, and to address puzzling questions regarding the apparent limited variety of meteorite sources. These exploratory encounters can also provide basic information that will facilitate the intelligent design of subsequent, more sophisticated asteroidal missions.

During the next decade it may also be expected that basic information regarding early solar system history will continue to flow from laboratory study of meteorites and stratospheric collection of interstellar particles. In this connection, it should be pointed out that to a large extent the current laboratory instrumentation used in this work was obtained in the course of lunar sample analysis during the 1960s and early 1970s. Attention must be given to modernizing the laboratories in which this work is done if this potentially major source of information is to be fully utilized.

Among the fundamental questions in the area of solar system origin are:

- *Formation of "single" stars.* Recent observations have shown that perhaps as many as 90 percent of all stars are members of binary systems; the Sun appears somewhat special in this regard. In order to specify the physical conditions in the earliest phases of the solar nebula, it is necessary to understand the way in which single stars are formed, how the process differs from that for formation of binary systems, and ways in which the formation of a planetary system is related to the way in which binary stellar systems are formed.

- *Nonstellar companions of the Sun.* The present paradigm for formation of the solar system involves a nebular component out of which the Sun forms. A fundamental question is whether other large, but nonstellar objects can form in the nebula at the same time that the Sun is forming. This question is of dual significance; it relates to whether there is a generic relationship between the formation of planetary and stellar binary systems, and it relates to the question of how protoplanetary objects may be formed. In this latter sense this question is a basic part of the question of how planets form.

- *Residual presolar disks.* In order to develop a clear understanding of the evolution of the solar nebula as that evolution affects the formation and evolution of individual components, it is important that we characterize the nebula and its properties as much as is possible. Particular emphasis should be placed on the mass and extent of the nebula, and its variation in time. Also of interest are the spatial and temporal variations of temperature and composition, both of which are essential to understanding the details of planet formation.

- *Formation of planetesimals.* One of the possible models for formation of the terrestrial planets, and perhaps for the cores of the outer planets as well, involves the formation of small solid bodies called planetesimals. Progress in understanding this pivotal phase, especially in the context of specific nebular models, is central to gaining an overall understanding of the planet formation problem.

- *Formation of the three major classes of planets.* Establishing the manner in which each of the three major classes of planets—terrestrial, jovian, and uranian—formed is a central issue in understanding the origin of the solar system. The temporal aspects of this topic are of particular importance; for example, if Jupiter formed before the other planets it may have controlled to some extent the formation of the other planets. If, on the other hand, the terrestrial planets formed before Jupiter, the details of their formation are likely to differ substantially from those associated with formation in the presence of a massive object like Jupiter.

- *Formation of satellites and small primitive bodies.* Much of the information that we have concerning the very early history of the solar system comes from the study of smaller bodies. The reason that the small bodies are significant in this regard is that they do not undergo much metamorphic evolution and therefore retain a reasonably unaltered record of conditions at the time of their formation. It is important then to place the formation of these objects in the context of broader nebular models.

- *Loss of the nebular gas.* One of the challenging problems in understanding the early evolution of the solar system is to identify when and how the bulk of the nebular gas (a hundredth of a solar mass or more) left the solar system. The timing is particularly relevant as it relates to whether accretion of solid material occurred in the presence or absence of significant amounts of gas. Probably at least Uranus and Neptune formed either after or during removal of the gas. In addition to understanding when removal occurred, it is important to understand the physical mechanisms involved in removing this amount of gas. There is growing evidence from astronomical studies that young stars that are thought to be similar to the Sun in its youth are associated with major bipolar mass loss. The relationship between this bipolar flow and dispersal of a circumstellar nebula is unknown.

- *Initial states of the planets.* A comprehensive study of the planets involves a characterization not only of their current state, but also of their evolution. An essential ingredient in this analysis is specification of the initial state of planets. Such specification includes temperature and compositional gradients or profiles. This type of information is provided through models of the formation of the planets and can be used in conjunction with observations to assess, for example, models for formation and evolution of planetary atmospheres.

Search for and Study of Other Planetary Systems

Underlying all modern efforts to understand the origin of the solar system is the assumption that the system is a natural result of star formation and that similar events take place when other stars are formed. A consequence of this view is that most stars should be accompanied by retinues of nonstellar companions. Empirical evidence about this relatively simple assumption and its broader consequences form the only viable foundation for significant advances in our understanding of the origin of the solar system.

A major challenge is to identify what features, if any, of the solar system are prototypical or representative of the general process of planetary system formation. For example, theoretical models are being developed that lead naturally to the expectation of the formation of jovian-mass planetary companions to all solar-type stars. Observational evidence either to validate this consequence or to refute it, and thereby provide an empirical foundation for a next generation of theoretical models, is thus essential to our knowledge of the origin of the solar system. Another fundamental challenge is to detect, study, and relate to theories of planetary formation the broad range of circumstellar material and phenomena that may represent precursor or post-formation phases of other planetary systems.

Anticipated State of Knowledge in 1995. At the present time there is no unambiguous evidence for the existence of another planetary system, let alone detailed information concerning statistical properties of planetary systems in general, or of structural details of specific planetary systems. IRAS and some ground-based instruments have detected discrete dusty structures associated with some nearby main-sequence stars, where the dust particles are some 2 orders of magnitude larger than typical interstellar dust particles ($\sim$100 μm versus $\sim$0.2 μm). However, these observations do not of themselves signify detection of another planetary system. Substellar companions to nearby stars have also been detected with the use of ground-based instrumentation. As yet, these observations are unable to distinguish between coplanar planetary systems similar to our own, and low mass members of more commonplace double star systems. Advances are under way to develop instruments that will permit a search of sufficient accuracy to allow a quantum jump in the available data base.

By 1995 we can expect to have 5 to 10 years of survey data from ground-based, special-purpose spectroscopic telescopes with detectors capable of measurements accurate to better than 10 m/s. While the spectroscopic technique of searching for other planetary systems does have its limitations, such a search will provide constraints on our current models. The ability of this technique to detect jovian-mass companions in orbits with periods of a few years is significant, and evidence either positive or negative on such objects would have a useful, perhaps revolutionary, effect on our efforts to understand the origin of the solar system. It is conceivable that another planetary system could be discovered prior to 1995 using some of the ground-based instrumentation that is currently being developed (e.g., high-precision spectroscopic and astrometric systems). It is very unlikely, however, that these systems will be able to go beyond a simple detection to provide significant data regarding the statistical properties of other planetary systems. It would appear that as of 1995, we will have only begun to collect the type of data essential to an understanding of the origin of the solar system. Detection of planetary systems therefore looms as one of the major challenges of any long-term plans in planetary research.

There are four major questions associated with this area of planetary research. The first concerns the possible uniqueness of our solar system. This can be answered by detection of at least one other planetary system. If this has not been accomplished prior to 1995, it should be a priority effort thereafter.

As exciting as the confirmed discovery of another planetary system would be, the major scientific advances of such a discovery will lie in answering the remaining three questions in this area of study. First, we must seek to characterize planetary systems on a statistical basis. For example: What is the frequency of occurrence of planetary systems as a function of spectral type of their central star? What are the masses of the planetary bodies relative to that of their central star? Answering these questions will require instrumentation capable of surveying a large number (100 to 1000) of stars to a level of accuracy that permits detection of objects comparable in mass to Uranus and Neptune. This type of data will permit us to deduce the general features of planetary systems—those features that must be explained by a comprehensive theory—from features that are peculiar to a given planetary system.

A third major question concerns the detailed nature of specific planetary systems. This type of effort must involve a longer time base as it requires precise determinations of planetary masses and orbits, and as much information on composition as possible. These types of data are required to both test and constrain models of processes within the gas-dust nebulae from whence planetary systems are presumed to form.

A fourth question concerns the cool material around stars in various stages of evolution. A central aspect of most modern theories on the formation of the solar system is the postulation of an accretion disk that evolves in such a way that it produces a central condensation (a star) and a suite of companions (planets). The previous questions focus on detection of the products of the disk evolution, the planetary bodies. No less important is an effort to characterize the nature and behavior of the disks associated with preplanetary systems, young planetary systems, and mature ones. Identification and study of this circumstellar material will help place the formation of planetary systems in the broader context of the problem of star formation.

These four major questions can be addressed provided that certain technologies and instrumentation are developed in a timely manner. The first three questions are most effectively addressed by an astrometric telescope of 10-μarcsec accuracy. Addressing the fourth question requires a low-scattering telescope to image and conduct spectrophotometric analysis of cool circumstellar material. Such systems must be located in space because turbulence in Earth's atmosphere places unacceptable limits on image size and stability at ground-based sites.

In this regard it is important to note that the manned base of the Space Station could provide an ideal platform for such a telescope. It has the long-term stability and data-handling capability required for a system of this type. A roughly 20-year observing program of all stars within 30 parsecs of the Sun is envisioned for this telescope. The technology is at hand to build such a telescope, and it would be a relatively inexpensive device by today's standards. This telescope and its scientific objectives provide perhaps the most exciting scientific use of the Space Station and should therefore be considered a prime candidate for implementation. It could complete its survey within 15 to 20 years of its initial operation.

Asteroids, Small Satellites, and Meteorites

General Characteristics

The asteroids are small bodies that orbit the Sun, for the most part at distances of 2 to 3.5 AU. The total mass of material in the asteroid belt is less than 0.001 earth masses. During the past 15 years there has been a phenomenal increase in our knowledge of the physical, chemical, and dynamical properties of asteroids. This has been achieved by earth-based telescopic observations of their positions, rotation, and spectral reflectance and emission.

The largest asteroid, 1 Ceres, has a diameter of 1000 km, and there are more than 30 others larger than 20 km. This population grades down to more numerous smaller bodies; there are 105 to 106 asteroids greater than 1 km in diameter. The size distribution is such that most of the mass is found in the largest bodies, however. The mean densities of asteroids seem to be 2 to 3 g/cm^3.

Optical properties, including variations of optical magnitude with rotation, optical and near-infrared spectral reflectance, and thermal radiance have been measured for a large number of asteroids. Most asteroids are distributed bimodally into one of two groups usually designated as C and S. Most abundant are the dark (geometric albedo of 0.03 to 0.04) C asteroids that exhibit rather featureless reflectance spectra. Both of these characteristics probably result from possibly small quantities of opaque minerals. This is consistent with identification of these asteroids as being of similar composition to carbonaceous meteorites, but the identification is not unique. Although there are marked exceptions (e.g., 313 Chaldaea), the C asteroids are concentrated toward the outer part of the asteroid belt.

The S asteroids exhibit larger geometric albedos ($\sim$0.16) and absorption features attributed to mixtures of pyroxene, olivine, and metal. Qualitatively, they are thus mineralogically similar to ordinary chondrites, as well as to fairly well-mixed igneous assemblages resulting from differentiation of asteroids of overall ordinary chondrite composition. Quantitatively, in most cases the relative proportions of these minerals do not appear to agree with those of ordinary chondrites. It is not clear whether this discrepancy is only a surficial phenomenon, or of greater significance. It is quite possible that some S asteroids are differentiated bodies, whereas others are similar or identical to ordinary chondrites in composition.

Some asteroidal reflectance spectra clearly indicate differentiation, such as E (enstatite) and M (metal). A number of asteroids, including many of the largest, do not fit into any of the standard classes. Some of these, particularly the 500-km-diameter body 4 Vesta, of basaltic composition, are presumably differentiated, whereas others, such as 1 Ceres and 2 Pallas may be undifferentiated.

The orbits of asteroids intersect one another, and the expected frequency of asteroidal collisions (relative velocity of ~ 5 km/s) can be calculated in a rather straightforward way. There are significant uncertainties in calculating the effects of these collisions, but it is very probable that almost all bodies less than 30 km in diameter are fragments of larger bodies, whereas asteroids larger than 100 km in diameter may be collisionally altered primordial planetesimals. The physical structure of all asteroids is undoubtedly influenced by their collisional history. In detail, these effects are not clear, the possibilities ranging from rubble piles of possibly bizarre geometrical appearance, to thick, blocky regoliths, to nearly clean rocky fragments. There is ambiguous evidence that some asteroids represent multiple, mutually orbiting bodies.

In the present solar system, almost all asteroids appear to be in orbits that are stable (except for the possibility of collision) on time scales that are long compared to the lifetime of the solar system ($\sim 10^{10}$ years). Exceptions are asteroids produced as collision fragments adjacent to regions in which their orbital period, or other characteristic frequencies, are in resonance with the major planets, particularly Jupiter.

The Earth-approaching Apollo and Amor asteroids are of some special interest. These bodies have perihelia within 1.3 AU, many of them pass inside the orbits of Earth and Venus, and a few pass within the orbit of Mercury. Close approaches to these planets cause the orbits of these bodies to be unstable on a time scale that is short compared to the age of the solar system. The observed Apollo-Amor objects thus constitute a steady-state population, and sources are required to sustain the observed population. These sources are probably twofold. Some bodies are fragments of main belt asteroids, transferred into Earth-approaching orbits by the same resonant mechanisms responsible for the delivery of some meteorites from the asteroid belt to Earth. Others are likely to be the devolatilized nuclei of short-period comets, the residue remaining after $\sim 10^4$ episodes of solar heating during perihelion passage. An uncertain estimate of the relative proportions, based on interpretation of observations and theoretical reasoning, is that at least 20 percent of these bodies are derived from the asteroid belt, and that at least 10 percent are of cometary origin.

The Earth-approaching bodies of asteroidal origin have a close kinship to meteorites, and many meteorites are probably collision fragments of these bodies. It is plausible to believe that a mission to a body of this kind would provide similar information to that obtained by a mission to a main belt asteroid of the same size but at a lower cost, because of the lower energy requirements characteristic of many of these near-Earth orbits. For example, the large Amor object 433 Eros ($\sim$20-km diameter) is likely to be of asteroidal origin. If so, it is undoubtedly a collision fragment rather than a primordial body, and may be expected to display the internal structure of a large asteroid. It is large enough to retain on its surface some of its collisional ejecta, facilitating collection of this material.

The Earth-approaching bodies most likely to be of cometary origin tend to be in higher-velocity orbits than those of asteroidal origin. It is not out of the question that some of the low-velocity bodies may also be of cometary origin, but there is no observational evidence supporting this at present. A mission to an extinct comet may be able to study and sample cometary material difficult to access during the active phase of a comet's life. Such a mission may also provide information regarding the possible relationship between comets and some carbonaceous meteorites. Although earth-based observations have resulted in remarkable progress in our understanding of asteroids, qualitatively new approaches will soon be required if this progress is to continue.

Anticipated State of Knowledge in 1995

Space Missions. The next major step in asteroidal science is clearly in situ studies of asteroidal bodies. Missions being planned for encounter prior to 1995, if carried out, provide a first step in this program. They represent two or three fast flybys of asteroids by spacecraft on the way to other prime target planetary objects. These include Galileo, CRAF (Comet Rendezvous/Asteroid Flyby), and the Saturn orbiter/Titan probe. These missions will be equipped with imaging that can achieve on the order of 100-m resolution of the asteroid surface. In addition, missions will map the surface in several spectral regions to provide mineralogical information. The observations should be of significant help in the interpretation of data from ground-based spectrophotometric observations limited to integral whole-disk reflectance spectra.

These missions will begin to address first-order questions such as large-scale topography, density, and composition of the asteroids. For the first time we will actually be able to see what an asteroid looks like, and will bring asteroidal observations to a state comparable to lunar studies prior to the space program.

The Voyager missions have imaged Amalthea and many of the smaller satellites of the Saturnian system, but at a resolution inadequate to tell much about the nature of the surfaces or the processes that have operated on these bodies. Phobos and Deimos are the only small satellites imaged with adequate resolution to tell much about the nature of the surface (5-m resolution was achieved by the Viking missions). The U.S.S.R. plans to make additional measurements of the martian satellites during the Phobos mission in 1988.

Ground-based, Earth-orbital, and Laboratory Studies. Important ground-based and laboratory studies of asteroids, minor satellites, and meteorites may be expected to continue during the next decade. Much of this work falls into the category of "small science," i.e., work that is best not programmed in advance, but allowed to evolve naturally through short-term, peer-reviewed grants.

It is obviously difficult to determine what we will have learned from these studies by 1995, but significant advances are likely. For example, the number of known asteroids, including those in near-Earth and Mars-crossing space, will grow substantially. A significantly increased number of near-Earth asteroids will be available for spectroscopic studies and comparison with the population of main-belt asteroids. The number of candidate near-Earth asteroids for future missions also will increase significantly.

Studies of asteroids by the Space Telescope (ST) also may advance our knowledge of the asteroids. The ST will provide images of about 20 km resolution at 2 AU from the Earth and will determine the shape of many asteroids. In addition, the high spatial resolution will allow reflectance spectra to be taken at many spots to test for surface heterogeneity. The capabilities of ST will also allow reflectance spectra to be taken in the ultraviolet region, and perhaps in the future, in more of the infrared region. These regions may show more diagnostic features than have been available from ground-based studies.

Many asteroids have been grouped according to spectral reflectance features, and many of these features have been related to known meteorite types. One difficulty in the link between asteroids and meteorites is that the reflectance spectra of the most common meteorites, the ordinary chondrites, are not precisely matched by the main-belt asteroid spectral reflectance groups. The closest match, some of the S asteroids, may actually be the source of the ordinary chondrites if the surficial features of the asteroids have been modified by exposure to the space environment. By 1995 these processes may be better understood as a result of laboratory studies, facilitating a closer observational link between the meteorites and asteroids.

There are good dynamical reasons for believing that most ordinary chondrite meteorites are derived from a limited (± 0.05 AU) region of the asteroid belt in the vicinity of the 3:1 Kirkwood gap at 2.5 AU. (Asteroids with values of their semimajor axes in this region will have periods in resonance with the orbital period of Jupiter). Except for the questions of spectrophotometric interpretation mentioned above, the known larger S asteroids, in the vicinity of 2.5 AU are prime candidate sources for ordinary chondrites. Most of the meteorite population is clearly derived from the main-belt asteroids, and their study provides our most detailed understanding of the nature and properties of the asteroids.

Because of the diversity of studies on meteorites, it is difficult to predict the state of knowledge in 1995. Prior to 1995, we can expect much new information will be obtained through the NSF-sponsored Antarctic meteorite collection program and the development of new analytical techniques. This collection represents a less-biased sample of the small meteorite end of the terrestrial meteorite flux and a substantial increase in the number of stony meteorites available for study. For this reason it may provide a look at meteorites derived from the asteroid belt at velocities so high that a significant yield of large meteorites is not expected.

This possibility appears to have already been realized by identification of several meteorites of lunar origin in the Antarctic collection.

A large group of meteorites, the achondrites, have clearly undergone igneous differentiation processes dated at 4.5 billion years—nearly contemporaneous with the formation of the solar system. Besides giving evidence that at least some asteroids were once geologically active, the meteorites allow the study of igneous processes that occurred under a set of conditions unlike those of Earth or Moon, and thus broaden our understanding of the effects of different parameters that cannot otherwise be studied in the laboratory. In addition, most of the iron meteorites formed very early in solar system history in cores of many (up to 50) different asteroids. These data provide strong evidence for the presence of a substantial heat source capable of producing primordial melting of the asteroids (and possibly planetary planetesimals as well). Many of the meteorites are breccias (made up of fragments of other preexisting rock) and provide evidence for the nature of collisional processes at an early epoch. The meteorites, and hence the asteroids, also retain a record of the cosmic-ray intensity in the past, as well as a record of significant (up to a few gauss) magnetic fields that were present before the asteroids accumulated. Several of the carbonaceous chondrites have been found to contain amino acids and other complicated organic molecules, which are clearly of extraterrestrial origin. Carbonaceous chondrites also contain calcium and aluminum-rich inclusions that appear to be among the earliest objects to have formed in our solar system. These inclusions provide evidence of very complicated processes in the early solar nebula involving multiple episodes of high temperature ($\sim$1400K). They also contain a record of the nucleosynthesis of elements that were added to our solar system just before the inclusions formed.

A problem associated with all of these meteorite studies, however, is the lack of a geological context for the interpretation of the observations. An asteroid sample return mission would go a long way toward providing this context. It would also provide the in situ data that would substantially increase the value of the remote studies of the asteroids. Similarly, the SNC meteorites can answer a number of questions about Mars, but they are clearly derived from a lava flow, whose composition is unlikely to represent that of the surface as a whole.

Questions in Asteroid Science. Accessing the information that asteroids contain about the early solar system will require much more detailed understanding of the asteroids' surface compositions, their internal structure, the degree to which they are heterogeneous, and their dynamical and collisional evolution.

At present our knowledge of the composition of specific asteroids is almost entirely founded on earth-based spectrophotometric data. Although we may expect this situation to change somewhat between now and 1995, for almost all asteroidal bodies it will still prevail. Earth-based data represent a surficial average over the entire observed disk of the asteroid, and heterogeneity is at present only crudely exhibited as a consequence of asteroid rotation. When looked at carefully enough, every asteroid appears spectrometrically unique, and different in detail from laboratory spectra of meteorites. Are these differences fundamental or are they primarily the result of heterogeneity or exposure to the surface environment or regolith phenomena? Making these distinctions requires remote sensing data of higher resolution. In addition, corroborating "ground truth" data are lacking, and will be required to ensure that unanticipated mineralogical differences are not overlooked or misinterpreted.

It is likely that asteroidal collisions have exposed interior regions of asteroidal bodies. When combined with higher resolution spectral and imaging data, this affords an opportunity to overcome the apparent limitation to surficial composition. This may facilitate addressing such questions as the fragmentation history and internal structure of differentiated asteroids, and the identification of specific asteroids as sources of particular meteorite classes.

A better understanding of this asteroid collisional evolution is also needed in order to learn which asteroids are primitive objects, as opposed to collision products of larger bodies, and whether they are best thought of as "rubble piles," megaregoliths, or simply as solid rocks. In this same connection, present observations of Hirayama families serve as a key source of information concerning asteroidal collision phenomena. Their apparent spectrophotometric heterogeneity calls into question the basic assumptions supporting these inferences.

With these more detailed compositional data it will also be possible to address the relationship between the heliocentric distance of asteroids and their chemical composition. Understanding this possible relationship is central to examining the conventional assumption that the formative solar system had a marked radial temperature dependence, related to the composition of planetesimals and planets formed at different locations in the early solar system.

COMETS

General Characteristics

Comets are thought to be small conglomerates of rock, ice, and dust—several kilometers in diameter and 10^{15} to 10^{18} g in mass—formed during the early years of the solar system's history. Today most reside in the so-called Oort cloud, in loosely bound orbits at tens of thousands of astronomical units from the Sun. Perturbations induced by the gravity of passing stars and interstellar clouds occasionally alter a comet's orbit and send it near the Sun, where solar heat evaporates the ice. The subsequent outflow of gas and entrained dust, illuminated by sunlight, produces the comets. Some of these comets that venture into the solar system are further influenced by the gravity of planets, becoming trapped in periodic orbits near comets. Some of these comets that venture into the solar system are further influenced by the gravity of planets, becoming trapped in periodic orbits near the Sun. Such periodic comets appear regularly for thousands of years until their volatile material is depleted.

Several schools of thought hold that comet-like objects were among the fundamental building blocks of some larger planetary bodies. As a result of their small sizes and their large average distances from the Sun, evolutionary processes that differentiated the planets are thought to have been insignificant for many comets. They may have played a role in later states of planetary evolution, perhaps by providing volatile constituents for some atmospheres. It has been speculated that some of these cometary constituents were essential to the origin of life.

A bright comet appears as a roughly spherical coma or atmosphere composed of comparable quantities of dust and volatile species such as neutral gases and ions. The curved, relatively featureless dust tail directed almost exactly away from the Sun shows considerable temporal and spatial structure. Depending on a comet's distance from the Sun and the light in which it is observed, its coma can be quite large, in the range of 10^4 to 10^7 km when a comet is at 1 AU. The plasma and dust tails are even larger in the case of a bright comet, some 10^7 to 10^8 km in extent. All of these phenomena result from the gas and dust that emanate from the nucleus; the nucleus itself is small, of the order of 1 to 10 km in diameter for typical comets.

The ice and snow in comet nuclei are composed of condensed gases and other volatile materials, including water and probably carbon monoxide, carbon dioxide, HCN, CH_3CN, and unidentified complex organic molecules. These species are the parent molecules of the molecules and ions observed in the coma and tail. The nonvolatile material is in the form of grains ranging from submicrometer-sized dust to sand grains and perhaps pebbles and boulders, containing silicates and possibly metals, oxides, sulfides, and organic compounds. When solar heat vaporizes the volatile material, the outflowing gas carries smaller solid particles with it. Since the comet's gravity is finite, though weak, any larger pebbles and boulders are likely to remain bound on the surface of the nucleus possibly leading to the formation of an "extinct" comet of asteroidal appearance. Some of the earth-approaching "asteroids" may be highly evolved comets of this kind. Since many comets show evidence of directional emission of gas and dust, it appears that the surfaces of their nuclei are inhomogeneous and may have localized active regions.

A tentative estimate of the volatile fractions of four recent comets has been made from their apparent production rates of carbon, oxygen, and nitrogen. Although the parent molecules are uncertain, they seem to be composed mainly of hydrogen, carbon, nitrogen, and oxygen. The mass ratio of the dust to gas liberated from a nucleus has been estimated in two cases to be 0.5 and 1.7 within a factor of 2. By comparison, the ratio of volatile to nonvolatile components is about 100 for solar material. These results imply that hydrogen and helium are depleted in comets. Nevertheless, comets seem to contain 3 to 10 times as much volatile material as the most volatile-rich meteorites. Thus comets appear to have formed from material at temperatures much lower than that characteristic of meteorites, at about 150K as opposed to more than 400K. This suggests that cometary material is the least-differentiated and best-preserved product of the preplanetary solar nebula that is known to remain in existence.

It is speculated that some fraction of comet dust may be unaltered interstellar material. It may be possible to illuminate this question by establishing some elemental isotopic ratios. Relative isotopic abundances of the elements reflect their formation processes. Carbon is a good example. Bodies within the solar system, including the Sun, Moon, terrestrial planets, meteorites, and Jupiter, exhibit a common value of about 90 for the $^{12}C/^{13}C$ isotopic ratio. Red giant stars show a range from 12 to 50 for this ratio. In carbon stars the ratio falls in a wide range from 2 to 100. Although observed values in the interstellar medium also span a wide range from 13 to 105, some investigators have argued that a value of 40 is representative. Many of the so-called "Brownlee particles," which are collected in the stratosphere, are suspected to originate in comets. At best, however, they are necessarily deprived of their most volatile components.

State of Knowledge in 1995

Our ideas about comets are constructed from limited remote observations as well as from our more general notions about solar system bodies. Significant advances in our knowledge of comets are expected during the decade preceding 1995, as a result of initial spacecraft missions carried out by a number of different space agencies. The NASA fly-through of Comet Giacobini-Zinner conducted the first in situ measurements of a comet. The instrument suite carried by that spacecraft was designed for solar wind and magnetospheric studies. The primary contribution of the mission was to characterize the particle and field distribution around the comet and to establish the basic features of that comet's interaction with the solar wind. Several important physical problems were addressed by the Giacobini-Zinner mission, including the medium-energy, nonthermal particle distribution, the morphology of the magnetic field, the character of the interaction between cometary gas and the solar wind plasma, and plasma instabilities excited by the interaction. Although the Giacobini-Zinner investigation was the first in situ comet investigation, neither the encounter orbit nor the instruments were chosen with the comet in mind. The same is true of the Pioneer Venus orbiter, which obtained images in radiation scattered by atomic hydrogen.

The three scientific spacecraft that encountered Halley's Comet were specifically designed for the purpose. The ESA Giotto project aimed for the closest approach to the comet's nucleus—passing as close as 500 km. Giotto passed through all of the outer comet/solar wind interaction layers and into the inner cometary coma. The instruments carried out a first-order characterization of the solar wind interaction, a crude characterization of the density and composition of cometary gases, and a crude analysis of the elemental composition of cometary dust. In addition, a limited number of medium-resolution pictures of the comet's nucleus were obtained.

The Soviet Halley project sent two spacecraft through the outer regions of the comet's interaction with the solar wind. Generally speaking, the information from the Soviet mission was similar to that from Giotto. The Japanese Suisei and Sakigake missions provided coordinated information on the solar wind flow upstream of the comet.

Altogether, the investigations of Comets Halley and Giacobini-Zinner achieved a gross characterization of the morphology of large-scale cometary phenomena. As valuable as these investigations were, most of the highest-priority questions that challenge our understanding of comets and that promise to reveal clues about the early solar system remain unanswered.

Detailed understanding of comets will require more intimate and extensive measurements than are accessible to flyby investigations. Comet rendezvous and comet nucleus sample return will be the means by which the major objectives of cometary science will be realized. The United States will carry out a comet rendezvous mission by the middle of the 1990s, which will follow a comet through most of its inner-solar-system passage. Analyses of cometary gas and dust, energetic particles, and magnetic fields and plasmas, as well as detailed investigations of the structure and gross composition of the comet nucleus will be carried out by the comet rendezvous mission.

Successful completion of the comet rendezvous should answer many outstanding questions about the gross characteristics of cometary features and phenomena. The next obvious step in the study of comets will be comet sample return. A crude sample return may be accomplished by flying a collector, at high velocity, through a cometary coma. However, the material returned in this way will retain only the information about its basic elemental and isotopic composition. In some ways, this situation resembles that of the stratospheric "Brownlee particles," which do in fact retain much of their original structure. Valuable as they are, these particles cannot be identified with any particular comet, and in fact may not even arise from comets at all.

By 1995, the next major step in cometary science will be to return an intact sample of comet nucleus material. This material can then be analyzed in detail in laboratories in order to carry out the mineralogical, chemical, and isotopic analyses that are needed to unravel the formation processes and evolutionary history of comets.

Questions in Cometary Science

It is expected that as a result of spacecraft missions and earth-based studies, our knowledge of comets will increase significantly during the next decade. It is clear that those results will represent only the beginning of our quest for understanding these most primitive and unevolved aggregates of matter assembled during the birth of the solar system.

An important set of questions concerns the description of the present state of cometary nuclei. These include descriptions of the chemical, isotopic, and mineralogical composition of comets, their internal structure, and the range of variation of these characteristics between different comets.

Questions of another class address cometary processes. As observed, comets are complex systems of neutral gas plasma, dust, and larger solid bodies. We must understand the physical processes that determine the loss of material from the cometary nucleus and the resulting short-term evolution of its structure, and those that produce the elaborate extended coma and tail of the comets.

Some of these same physical processes determine the nongravitational evolution of cometary orbits, that is, a "rocket effect" caused by the asymmetrical emission of gases from the nucleus. These altered orbits can have a substantial effect on the probability that a comet will impact planets and are central to the question of the evolution of active comets into Apollo objects of asteroidal appearance.

This knowledge of the present state of comets and the physical processes to which they are subject is required to understand the properties of earth-impacting material derived from comets. It is known that meteors, meteorites, and cosmic dust represent the impact of cometary and asteroidal materials, but the relationship of this material to their sources is imperfectly understood. Like returned samples, laboratory study of this material provides important information regarding these bodies and their origin. Characterization and identification of possible cometary material are required.

The resulting knowledge of present-day cometary composition, structure, and processes is fundamental to understanding where, when, and how the comet nuclei formed. In particular, we need to understand the age and any alteration of the various components of cometary nuclei. To what extent do they represent evolved solar system material, early solar nebula condensates, or solids of interstellar origin? This extrapolation back to the time of the origin of the solar system is also needed to interpret the extent to which comet-like objects contributed to the formation of the giant planets and the volatile inventories of the terrestrial planets.

Comet Measurements and Technical Requirements

Addressing the scientific questions cited above will require detailed investigations of the comet's nucleus and the emitted dust, as well as the gas, plasmas, and fields in the comet.

Cometary Nucleus. Investigations of the comet nucleus should be aimed at establishing its composition and its physical and structural characteristics. Compositional measurements should determine the atomic, molecular, and mineralogical content of the refractory and the volatile solids. Together with measurements of the physical and structural features, this information will help in ascertaining the history of cometary matter and the processes responsible for its formation and for the assembly of cometary nuclei.

Accurate isotopic measurements should be carried out on both the refractory and volatile constituents to explore the nucleosynthetic history and establish a time scale for major events in the history of cometary material. Both the compositional and structural investigations should be extended over the variety of physical scales that characterize cometary nucleus material; this covers a range from the microscopic grains to the full size of major macroscopic components of cometary nuclei. It is desirable to identify the major mineral assemblages of the nucleus for those constituents that make up 5 percent or more of the comet's composition, and with a resolution of better than 10 percent of the nuclear diameter. On the small scales, complete determination of the structure and composition of cometary material requires detailed analysis of the dust component to discern the dust's character and origin.

Cometary Atmosphere. Measurements of the gaseous component of cometary effluent will reveal information about the most volatile materials in the nucleus. Spacecraft should have the capability to identify and determine the abundances of all molecular species in the mass range 1 to several hundred, and to determine the isotopic ratios for the important species at an accuracy sufficient to identify deviations from solar composition and other anomalous isotopic variations. A major purpose of the cometary atmospheric measurements is to ascertain the composition of the so-called parent molecules of cometary effluent and their evolution as they leave the comet. Both the neutral and ionized components of the cometary atmosphere should be analyzed in detail, including the variations in composition with distance from the comet's nucleus.

Solar Wind Interaction. The structure of the large-scale cometary phenomena should be determined. Measurements of the electromagnetic fields, plasma, energetic particles, and neutral gas should be made in a volume surrounding the nucleus and encompassing the upstream coma and solar wind interaction region, as well as a significant volume of the tail. Measurements should be made with sufficient spatial and temporal completeness to allow identification of the major dynamical physical processes, including transients, that play important roles in shaping the overall cometary structure and providing its sources of energy.

Technical Capabilities. Cometary studies require a sophisticated complement of spaceborne and laboratory instrumentation that will not be discussed in detail here. Scientific investigations of comets need a launch system able to reach a variety of orbits and to maneuver for an extended period of time near the comet. At present, low-thrust electric propulsion seems to provide, by a wide margin, the best propulsion system for such missions.

Achieving the scientific objectives of comet exploration will require missions of months' duration designed to carry out extended investigations of a single comet during both its most active and its more quiescent times. Return of high-integrity samples of cometary nucleus material to Earth for detailed analysis in terrestrial laboratories will be essential to realize the overall goals of comet science, as well as of solar system science in general. However, adequate analyses should be carried out in situ, because these materials may well undergo significant changes once they are removed from their natural environment.

4
Future Programs

PROPOSED MISSIONS

Programs for Planetary Geosciences

Types of Missions

The goals of planetary exploration are met through observations and missions in which the levels of investigation are generally progressive. Earth-based observations provide limited, but important data that allow the formulation of first-order questions.

As the first level of investigation, reconnaissance by flyby missions attempts to reveal the major characteristics of a planet, such as its radius, mass, rotation rate, and the existence of magnetic fields, an atmosphere, oceans, satellites, and the like. The exploration phase follows and has the goal of describing and understanding the state of a planet and the general processes that have influenced its environment. This phase is carried out by long-lived orbiters equipped with a variety of cameras and other remote sensing instruments and may include entry probes to measure the chemical composition of the planet's atmosphere or surface. Such missions can image the surface of the planet; provide a global map of the distribution of elements and minerals on the planet's surface;

determine a planet's global properties including topography, gravity, and magnetic fields and density distribution; and characterize the atmospheric and ionospheric structure and dynamics.

The intensive phase of investigation addresses the highest-order questions revealed by the earlier phases and involves sophisticated and complex missions. Detailed study of the properties of surface materials, interaction of surface and atmosphere, stratigraphic and depositional history, and biologic questions requires in situ measurements by soft-landed automated laboratories, mobile laboratories (rovers), and networks of instruments.

"Network science" is defined as geophysical measurements made over a relatively long time (>1 earth year) at several locations over a planet's or satellite's surface. Although network science can be accomplished with a minimum of 3 to 4 stations, ideally a system would include an array of 6 to 12 stations with instruments to measure seismic events, physical properties of the surface, heat flow, and (where applicable) meteorological parameters. The stations must provide simultaneous measurements from the seismic and meteorological experiments, and may also make elemental chemical and mineralogic measurements with appropriate instruments. Network science can address geophysical questions on local, regional, or global scales. For example, most knowledge of the interior of the Earth has been derived from seismic data. Networks of seismometers on a global scale yield information about the properties of the core, mantle, and lithosphere; arrays of seismometers on a regional scale can provide detail about systems of active faults; and local arrays (tens of kilometers or less) can provide information on the presence and configuration of rock layers within the lithosphere. Network stations can be emplaced by penetrators, semi-hard or soft landers, or a rover.

The measurement of heat flow is fundamental for understanding the interior characteristics of planets and satellites, yet obtaining valid data remains a major technological problem. All present means for emplacing instruments in the subsurface (e.g., penetrators, drilled holes) disturb the thermal regime for a period of time that exceeds the lifetime of most missions. These problems must be overcome, or alternative means found for measuring heat flow.

Sample return missions can provide fundamental data that can be acquired in no other way about the composition, history, and evolution of other worlds. At the same time, sample return missions present exciting technical challenges; they require sophisticated robotic spacecraft systems and complex operational capabilities on planetary surfaces.

Both terrestrial and extraterrestrial rocks preserve evidence of past events, thereby providing essential clues to understanding the planet's origin, history, and evolution. The fabric of crystals that compose a rock (or a piece of solid ice from a comet) reflects both the original formation and subsequent evolution. One can distinguish between lava flows that cooled quickly at the surface and deep-seated crustal rocks that cooled slowly tens or hundreds of kilometers down. The different minerals in a rock reflect further details of their formation—the formation temperature, the cooling rate, the nature and abundance of volatiles, and the genetic relations between minerals formed at different times. The detailed chemical character of a rock can thus be the key to identifying global planetary processes—core formation, crustal separation, or episodes of widespread volcanism. The measurement of radioactive parent and daughter elements in a rock provides independent information about the timing of major planetary events—origin, major meteorite impacts, and volcanism.

Analyses and studies of samples returned to the Earth are unique in that they: (1) can be performed by a variety of scientists with current state-of-the-art technology; (2) permit iterative, imaginative experiments that can be based on prior results, including unexpected ones; (3) allow effective separation and concentration of mineral phases, based on the specific properties of the sample; (4) permit many different analyses on the same sample; and (5) permit the deferral of certain experiments, if necessary, until better analytical technology or understanding is available. The flexibility of laboratory sample analyses, and the resulting confidence in results, is largely due to the high analytical precision and to the greater control of experimental parameters possible in a laboratory.

Studies of the mineralogy, mineral chemistry, texture, and bulk chemical composition will define the physical and chemical history of rocks. Evidence for processes ranging from crustal formation to volcanism or chemical weathering at the surface will be addressed through detailed comparison of textural properties, mineral composition, and the distribution of elements and isotopes within the rocks.

A wide variety of signatures have been identified among trace elements as tracers for terrestrial and lunar geochemical processes. Analyses of these various types of elements, either in groups or as pairs, will yield evidence on the nature of the bulk starting materials with regard to differentiation, the degree of that differentiation, the internal heat sources, the temperatures and pressures of internal processes, and the nature of meteoritic material impacting the body during the geological past. Experience shows that these analyses can be combined with other geological information to unravel the complex evolutionary history of planetary surface materials.

Precise isotopic analyses will allow us to solve a wide variety of chronologic and geochemical problems. Long-lived radioactive species and their products (U-Th-Pb, K-Ar, Rb-Sr, Nd-Sm) provide isotopic ages for rocks and are the only means of establishing an absolute chronology. Stable isotopes (H, O, N, Si, C, S) provide very powerful geochemical tracers that can be used with the chronological data to explore past states of the interior as well as more recent surface processes. Anomalies left by the decay of extinct, short-lived radioactive isotopes can be expected to provide evidence for preaccretion conditions and time scales in the early solar system, before formation of the planets.

Analyses for He, Ne, Ar, Kr, Xe, and their isotopes are essential for understanding the internal differentiation history, its interaction with cosmic radiation, and the evolution of its atmosphere. Such data from surface materials will provide powerful tools for understanding the evolution of the planet's surface and interior. From our experience with lunar samples, meteorites, and terrestrial rocks, Xe isotopes are expected to be the most versatile, as the isotopic patterns may reflect several processes—extinct short-lived isotopes, fission of long-lived and extinct isotopes of U and Pu, and the mixing effects of various reservoirs of gas.

Samples will be examined for evidence of remanent magnetization related to any past magnetic fields, as well as for a variety of physical properties, such as grain distribution, density, porosity, thermal conductivity, capacity for retaining volatiles, and seismic wave velocity. These measurements will provide basic data for various physical models.

Planned Missions

The missions proposed to study the solar system during the period 1995 to 2015 follow the balanced approach recommended by the National Research Council's Committee on Planetary and Lunar Exploration (COMPLEX). By 1995, the exploratory and reconnaissance phases should have been completed for the inner planets, with the important exception of Mercury, and the proposed missions will involve the intensive study phase. The Voyager spacecraft will have performed the initial reconnaissance study of the satellites of Jupiter in 1979, of Saturn in 1980-1981, of Uranus in 1986, and of Neptune in 1989; the Galileo spacecraft will have continued exploration of Jupiter's satellites in the 1990s.

During the early part of the study period, the task group recommends that Galileo-like missions explore the Saturn, Uranus, and Neptune systems. These spacecraft, with an extensive array of remote-sensing instruments, would carry out repeated flybys to characterize the major satellites. Because of scientific interest in its atmosphere, Titan would receive more intensive study, with radar investigations and a probe to measure the atmospheric composition.

Our own Moon remains a body of the highest scientific importance. The task group recommends that the global survey by the Lunar Geoscience Orbiter be followed by deployment of rovers and a geophysical network, as well as resumption of sample return from selected locations.

The task group also recommends a Mercury orbiter, with a landed transponder if possible, to complete the basic characterization of the inner planets. Mercury's composition, mass, and magnetic field provide key tests to many theories on the evolution of the solar system. The mission will not only involve planetary science, but will include solar studies and some tests of relativity physics by careful tracking of the planet's motion. The discovery of new trajectories has made this mission possible with existing propulsion systems.

The major mission recommended for the initial decade of the study period is a Mars rover and sample return program consisting of linked missions launched in 1996 to 1998. This is envisioned as a very comprehensive mission—one with a capable rover to collect selected samples for return to Earth and carry out extensive observations over the surface of the planet. The return of unsterilized martian materials could provide unique data on the absolute chronology of martian rock units, on detailed detection and characterization of contemporary or fossilized life, on surface-atmospheric interaction processes and rates, and on the composition and evolution of Mars' crust and mantle. A sample return should, if possible, have the capability to provide rationally chosen samples from a number of carefully selected areas in order to maximize the value of the samples.

Study of returned martian samples on Earth provides an excellent opportunity to look for any evidence of martian life, past or present. To protect the geochemical and biological integrity of the returned sample, sterilization by any method must be avoided. Because any martian organisms included with the returned sample might be killed by exposure to the high pressure, high water content, and high oxygen content of the terrestrial atmosphere, the most promising life detection "experiments" may be those based on chemistry and morphology, rather than on metabolism. Such experiments would include "micropaleontology" examinations, perhaps using stains that are reactive with carbon compounds. If any living systems should be detected in the returned sample—whether viable, dormant, recently dead, or fossil—the direction of our future exploration of Mars would be completely changed and there might be an early reexamination of manned missions to the planet.

A rover is necessary as a mobile sampling device in order to ensure that a wide enough variety of samples is collected to meet the mission objectives. The rover should have those capabilities necessary for sample collection, examination, and characterization. In general, the rover capabilities needed are roughly those of a human geologist collecting samples in the field. Like a geologist, the rover should obtain multispectral, stereoscopic images at a variety of scales and resolutions, process them, and interpret them by comparison to images derived from previous experience. It should lift samples to examine their details closely and to estimate their weight and density, thereby evaluating the amount of weathering. It should carry out simple chemical tests analogous to those made with a geologist's traditional Geiger counter and acid bottle. It should provide this information to Earth so that a decision can be made either to collect the sample or to discard it and move on to another.

Although the rover's prime objective is to support sample collection, it is important to note that its capabilities, and the data it collects to characterize possible samples, would also be scientifically important during an extended traverse to analyze and characterize martian surface materials up to a significant distance from the landing site. After the rover has completed its sampling traverses (first in the vicinity of the landing site and then, if possible, at greater distances), it could carry out an important and exciting surface traverse of Mars, making the same observations over long distances. Such a post-sampling traverse would provide an important regional context for the sample suite and would also help understand better the complex processes that have taken place on the martian surface. Eventually, manned missions will offer the most complete and comprehensive execution of the intensive phase of planetary exploration.

During the latter part of the study period the task group proposes to continue geoscience investigations with ongoing investigation of Mars as well as with a number of missions that are now less well defined and of somewhat lower priority, or that will require new technological developments before they can be carried out. It is important to deploy a network of stations on the surface of Mars to study its interior structure with seismic techniques and to understand the global meteorology. Sensor networks should eventually be emplaced on all the inner planets for continued seismic and other studies over a long period of time. Venus has the highest priority, but the high surface temperatures will make this mission very difficult.

Detailed study of returned samples will continue to have high priority throughout the study period. These studies will include samples from new locations on Mars, including the polar regions, and on the Moon, including the far side. Eventually, samples should be returned from Mercury and Venus.

A mission of special importance is the investigation of the "surface" of Titan, but the nature of this surface—solid or liquid—will not be known until radar investigations have been made from a spacecraft. A similarly important mission is a lander on Io to investigate the composition of the materials emitted by volcanoes, both active and inactive.

Programs for the Outer Solar System

Types of Missions

Addressing the goals set forth in Chapter 2 will demand new missions. Required missions to the outer solar system include long-lived orbiters, atmospheric probes, deep atmospheric probes, and a ring rendezvous spacecraft. These new missions will build on the first look provided by Voyager, which will have observed all the giant planets: Jupiter, Saturn, Uranus, and Neptune.

Orbiters provide multiple looks for remote sensing of planetary surfaces and atmospheres, and for direct measurements of the magnetosphere at many locations. They also provide a long time scale to study dynamic phenomena. New discoveries can be investigated in greater depth by changing the later stages of an orbiter mission to respond to new findings.

Although remote sensing has important virtues of its own, some kinds of measurement can only be made from within an atmosphere, and others can be made much more accurately there. Notable examples are the abundances of noble gases—the principal clue to the possible presence and nature of a primary atmosphere—and the abundance of nitrogen, which can be remotely measured only under favorable circumstances. In general, in situ abundance measurements can be much more accurate than those made remotely, and have given us most of our current isotopic data. Further, a descending probe can be tracked to give a vertical wind profile, whereas remote tracking of clouds gives a nearly global view, but only at one or a very few heights. Measurement of cloud properties and radiation balance are also best carried out from a descending probe.

Special techniques will be required to study Saturn's rings directly. A proposed ring rendezvous mission would use low-thrust propulsion to orbit Saturn in a plane just above the equatorial (ring) plane. By slowly decreasing the semimajor axis and altitude above the ring plane, a spacecraft could make multiple, low-relative-velocity encounters with ring particles. This would allow direct measurement of particle physical and chemical properties, and direct observations of inter-particle collisions.

Planned Missions

As described in the recent publication, *A Strategy for Exploration of the Outer Planets: 1986-1996* (National Academy Press, 1986), the most immediate priority is intensive study of the Saturn system. To address the scientific questions concerning this system will require a long-lived orbiter and probe investigation of the atmosphere of Titan, and perhaps Saturn as well. The orbiter would study the planetary atmosphere, the magnetosphere, the rings, the surfaces of the satellites, and the surface of Titan by radar. The probe would be similar to the Galileo probe, making in situ measurements and perhaps images. Important scientific objectives of this mission are to determine the composition and structure of the atmospheres of Saturn and Titan; make detailed, long-term studies of Saturn's rings; investigate Saturn's small satellites and the magnetosphere; and measure the physical nature and state of Titan's surface.

In addition to the Saturn system, the Uranus and Neptune systems will also become important subjects for study after 1995. Prior to 1995, the Voyager 2 spacecraft will have made a preliminary reconnaissance of these two systems. This will provide a basis for planning future intensive studies of these planets. The task group expects that many of the same questions about their atmospheres, magnetospheres, satellites, and rings may arise. This will lead to space missions similar to Galileo and the Saturn orbiter with Titan probe. An orbiter will provide exploration of the system, multiple looks at the planet and its satellites, and a long time base for study of meteorology, ring dynamics, and magnetospheric interactions. The probe will make direct measurements of the composition, cloudiness, and vertical structure of the planetary atmosphere (or perhaps the atmosphere of a satellite).

Future Missions and Programs for Primitive Bodies and the Origin of the Solar System

We can expect that, by 1995, reconnaissance missions to comet nuclei and asteroids will have transformed our view of these bodies from unresolvable points of light into planetary bodies of distinct shape and surface morphology. It is hoped that a good start will have been made in understanding their chemical and mineralogical structure, and their relationship to the small fragments of these bodies sampled on Earth in the form of meteorites and interplanetary dust. Making full use of the potential information obtainable from these bodies will require initiation of a new phase of detailed study.

Multiple flyby and rendezvous missions to asteroids will be needed to address questions of the variety and spatial distribution of asteroidal bodies. In situ data must be related to earth-based spectrophotometric data, and to physical theories of asteroid collisional fragmentation and evolution into earth-crossing meteoritic fragments. Such missions can also identify the range of morphological and structural characteristics of asteroids, and permit addressing the question of separating the record of early solar system history from subsequent collisional processing. With the background and understanding obtained from such studies, it will be possible to wisely select samples for return to laboratories on Earth for detailed chemical, mineralogical, and isotopic investigations of asteroidal material. Measurements of this kind are not only essential to interpretation of early solar system events in a context of planetary geology, but also may provide the key for similar interpretation of the large quantity of data obtainable from meteorites.

The study of comets is in a way easier than that of asteroids, because cometary activity releases large quantities of material from the nucleus into the atmosphere from which it can be sampled without actually landing and collecting samples from the surface. Valuable chemical and isotopic information can be obtained from a returned sample collected in this way, even if some mineralogical structure of the material is destroyed during the collection. A sample return mission of this kind could be of much importance in itself, as well as a valuable forerunner of collection and return to Earth of "pristine" samples of a cometary nucleus.

The earth-approaching Apollo and Amor objects are a special class of primitive bodies. Some of these are likely to be ~1-km-diameter asteroidal fragments derived from main-belt asteroids and transferred to near-Earth orbits by the same resonance mechanisms responsible for the transfer of smaller meteorite-size fragments from the asteroid belt. In fact, objects of this kind are likely to be the more immediate parent bodies of many of the meteorites in our collections. Understanding the physical structure of this fragmented material should be of much value in understanding the evolution of the steady-state collisional size hierarchy and the effects of this collisional history on the natural sampling of asteroidal material in the form of meteorites.

Other Apollo-Amor objects exhibit orbits and physical characteristics suggesting they are likely to be the devolatilized residue of cometary cores. These can therefore provide an opportunity to sample cometary material that may not be readily available at the surface of an active comet. By displaying the end product of cometary evolution, these bodies can provide better understanding of the active processes of comets, and thereby facilitate interpretation of remote observational data on comets.

The task of developing the instrumentation and technical means of successfully effecting these future measurements and remote sample returns will be a challenge. The variety and number of rendezvous and sample return opportunities to primitive bodies are now limited by existing ballistic propulsion systems. Full exploitation of the potential offered by such studies will require development of low-thrust propulsion systems that permit selection of targets primarily because of their scientific interest rather than on the basis of their accessibility.

As discussed in Chapter 3, the fundamental question of the origin of the solar system will not be answered simply by future missions, but will require highly disciplined yet imaginative integration of theory and observation. Adequate support of such investigations, including availability of advanced computing systems, will be needed for this work to proceed apace. Investigations of this kind will depend heavily on the understanding obtained from studies of primitive bodies. However, in the long term, evidence for the existence and characteristics of other planetary systems should prove to be of comparable, or even greater importance to resolving this ancient and fundamental subject of human thought. It is essential that we conduct a comprehensive search for other planetary systems.

A variety of observational techniques can and should be brought to bear on this problem. Before 1995, ground-based spectroscopic and astrometric studies will provide a preliminary survey of some of the nearby stars that may harbor planetary systems we can detect. The Hubble Space Telescope can be expected to make important contributions in this area, as can the Space Infrared Telescope Facility. A comprehensive search for and study of other planetary systems will, however, require an astrometric telescope in earth orbit. The need to have a long-term (10 to 20 years) systematic observational program indicates that the telescope should be on the Space Station. If this activity is initiated in 1995, at the beginning of the era under consideration, a full survey should be complete by 2015.

Results from this type of survey, which needs to be conducted with a system capable of 5- to 10-arcsec accuracy in terms of its ability to determine the relative positions of stars, would be available continuously over the two decades of observing indicated for this program. Constructing such an instrument is the principal new activity that needs to be initiated under a general program to search for other planetary systems. We can expect that as results come in from this Space Station Astrometric Facility, they will suggest further instrumentation needed to search for and interpret other planetary systems.

A PROGRAM FOR INTENSIVE EXPLORATION OF MARS

Concurrent with the continuing exploration of the solar system, the Task Group on Planetary Exploration recommends a special focus on the planet Mars because of the unique technical and scientific opportunity it provides. This campaign to understand Mars would proceed in phases, starting with missions that advance our current understanding, and extending to the eventual use of exploratory capabilities on the Mars surface currently possible only for humans. This focus on a planet similar to Earth will provide dividends in knowledge of Mars, of Earth, and of the origin and evolution of the planetary system. For reasons of its ease of exploration, comparability to the Earth and Moon, geologically active history, record of climate change, and possible environment for origin of life, Mars is unquestionably the best planet on which to focus this campaign. Already we have enough detailed information on Mars to identify sites where intensive study of the surface would provide critical information on the crustal evolution, geologic history, present and past water and atmosphere inventory, and history of climate variation. We recommend the timely inception of this campaign to understand Mars, in concert with the continuing exploration of the remainder of the solar system.

Why Mars?

Two of the major goals that motivate intensive investigations of the solar system are to understand the histories of planets and the processes that dominate their evolution and control their environment, and to learn what conditions and mechanisms were responsible for the origin and evolution of biological systems. The terrestrial planets occupy a special place in solar system investigations with respect to these goals.

Earth is a terrestrial planet; advances in our understanding of the Earth and its environment are inextricably tied to advances in our understanding of terrestrial planets as a class. Moreover, the future of human life in the solar system depends on the continuing quality of Earth's environment, as well as on the possible spread of life to other planets. Human activity on Earth is beginning to have a significant influence on its environment. There is a concomitant need to understand and be able to predict the environmental consequences of human activity while control is still possible.

Earth, Mars, and Venus constitute the triad of basically similar terrestrial planets that have undergone evolutionary processes culminating in major environmental changes. These environmental changes, particularly in the case of Mars, challenge our understanding of the behavior of the terrestrial environment.

The study of the terrestrial planets as a class will bring a depth of understanding not otherwise achievable. This general understanding will have direct application to the Earth in particular. A scientific understanding of planetary phenomena requires investigation of similar phenomena over a range of physical conditions. By making comparisons among the terrestrial planets and by deriving theories capable of correctly encompassing observations on the several bodies, we will gain an understanding that is not encumbered by ideas tested only against the conditions existing on Earth.

Because of the interest that they hold within the broader context of solar system studies, the terrestrial planets should be a special concern of solar system exploration within an overall balanced program. This general conclusion leads to a recommendation for a detailed exploration of the most available terrestrial planet, Mars.

Mars is unquestionably easier to explore than Venus: the surface temperature and pressure are within the range where machines and humans can operate for long periods. Except for occasional global dust storms, the Mars surface is not obscured by its atmosphere, allowing remote measurements both to select and monitor sites of intensive exploration.

Mars can be compared fruitfully with the other terrestrial planets in several areas. It formed further from the Sun than the Earth, and this provides a test for models of planetary formation. Its different size and composition can test models for planetary accretion and evolution. The bulk composition of Mars, necessary to test models of planetary formation, is significantly constrained because we do not know the dimensions and composition of the core and mantle. The history of accretion and differentiation of Mars could be clarified by a better understanding of the ages and distribution of components in the crust and atmosphere.

The evolution of Mars' crust differs from both the Moon's and the Earth's. On the Moon, the primitive crust was certainly dry and may have arisen from an early magma ocean. Subsequently, little crustal evolution has occurred other than thin mare basalt flows. On the Earth, no primitive crust survives. Its current crust falls between two extreme types: continental, formed in a wet environment 2.5 to 3.5 billion years ago, and oceanic, formed by basalts derived from the upper mantle, less than 200 million years ago. The primitive crust on Mars is thought to have formed in a volatile-rich environment, but the details of its composition are not known. Extensive volcanism has occurred throughout its history.

Like the Earth, Mars appears to have a global dichotomy in its crust. On Mars the southern hemisphere is old and cratered, whereas the low-lying northern hemisphere is dominated by younger flows and other deposits. To understand this dichotomy we need to know the nature, cause, and timing of the geological processes involved in forming the terrains, and we need to understand the characteristics of the complex boundary between them.

Like the Earth, Mars may have continuing geologic activity. Giant volcanoes, great canyons, multiple lava flows, and polar ice caps are all present on Mars. Each has analogues on the Earth. However, one major difference between the planets is the lack of global plate tectonics on Mars. Recognition of the process of seafloor spreading and plate tectonics was a breakthrough that provided a unification of much diverse knowledge of Earth's history. The martian environment, being distinctly different, will provide additional constraints on our understanding of the geologic history of the Earth.

Mars has clearly experienced major climate change over its history. The early presence of liquid water and a substantial atmosphere is shown by the weathering of old craters and the dendritic stream channels on ancient surfaces. Further, a record of at least recent climate change may be preserved in the sedimentary layered rocks near the martian poles. The variation in orbital parameters, which may drive the glacial stages on the Earth, is much more extreme on Mars. This combination of more extreme orbital variation and the record of the atmospheric response in the polar laminar terrain is a compelling reason to study Mars' climate change in detail. Further knowledge would provide models for atmospheric evolution under a distinctly different situation from Earth.

Accumulating geological and climatological evidence suggests that early Mars may have been suitable for the development of life during the first billion years of its history. Earth and Mars may have been quite similar during that period when algae flourished on Earth. Although the conditions on Mars currently are very difficult for the survival of life forms as we know them, and although the Viking missions found no evidence for life or organic material, the possibility of finding traces of past life provides a scientific objective for more detailed study. The Mars near-surface rock layers possibly could contain microfossils remaining from this early period, or records of chemical change on the earlier surfaces caused by early life. Sedimentary deposits that date back to the first billion years in Mars' history may be exposed in the walls of the great equatorial canyons. Analysis of the geologic history and detailed study of such deposits would have a significant impact on our understanding of the origin of life.

Scientific Objectives for a Mars Focus

The scientific utility of studying a second terrestrial planet in sufficient depth that the past and present processes can be identified and compared with similar ones on the Earth leads naturally to the recommendation of a major focus on intensive exploration of Mars. Current understanding of the origin, early history, and present state of Mars motivates a set of scientific goals whose accomplishment defines the scope of the recommended campaign to understand Mars:

1. Characterize the internal structure, dynamics, and bulk composition of the planet.

2. Characterize the chemical composition, structural features, and mineralogy of surface materials on a regional and global scale.

3. Determine the chemical composition, mineralogy, and absolute ages of rocks and soil for the principal geologic provinces.

4. Characterize the processes that have produced the landforms of the planet.

5. Determine the chemical and isotopic composition, distribution, and transport of volatile compounds that relate to the

formation and chemical evolution of the atmosphere, and their incorporation in surface rocks and polar ice.

6. Characterize the planetary magnetic field and its interaction with the upper atmosphere, solar radiation, and the solar wind.

7. Determine the extent of organic chemical and possible biological evolution of Mars, and explain how the history of the planet constrains these evolutionary processes.

Some of these objectives can be addressed by NASA's planned missions, including the Mars Observer and the Mars Aeronomy Orbiter. These will provide a global map of the distribution of elements and possibly minerals on its surface; determine its global properties including topography, gravity, and magnetic fields; and characterize the atmospheric and ionospheric structure and dynamics. However, more sophisticated and complex missions are required to answer the higher order questions. The necessity for detailed study of the properties of surface materials, interaction of surface and atmosphere, and stratigraphic and depositional history, and the need for addressing biological questions require in situ measurements by complex, automated laboratories or rovers, and networks of instruments.

The measurement of heat flow is fundamental for understanding the interior characteristics of planets and satellites; yet obtaining valid data remains a major technological problem. All present means for emplacing instruments in the subsurface (e.g., penetrators, drilled holes) disturb the thermal regime for a period of time that exceeds the lifetime of most missions. These problems must be overcome, or alternative means found for measuring heat flow.

The next major scientific objectives beyond those addressed by current missions could be met with a capable rover to collect selected samples for return to Earth and to carry out extensive observations over the surface of the planet. The return and study of pristine martian materials could provide data on the absolute chronology of martian rock units, on detailed detection and characterization of possible contemporary or fossilized life, on surface-atmospheric interaction processes and rates, and on the composition and evolution of Mars' crust and mantle. A sample return mission should provide rationally chosen samples from a number of carefully selected areas in order to maximize the value of the samples.

Detailed study of returned samples will continue to have high priority long after their initial return. These samples should be from many carefully chosen locations on Mars, including the polar regions. The collection of these samples involves increased complexity with far-traveling, highly capable rover laboratory/observers and drill stations capable of drilling deep holes. Instruments to measure seismic waves, composition and physical properties of the surface, heat flow, and meteorological parameters should be emplaced in a network of at least 3 to 4 stations; 6 to 12 stations would be preferable.

Potential landing sites can be selected from high-resolution Viking orbiter mosaics. Along with previous knowledge, this information allows the identification of locations where future surface exploration would provide substantial scientific return. Several interesting sites have been identified that contain a rich diversity of geologic ages and rock types.

Volcanic deposits are exposed in the southeast portion of the scarp surrounding Olympus Mons. Evidence of tectonic activity is visible in the layered deposits in Candor Chasma and in the mesas at the bottom of the fault trough. Polar layered deposits, which may represent climatic change, are available near the north polar cap, which is, itself, of great interest. A wide variety of geologic features are available in the Mangala Valley, including stream deposits and very young lava flows. For each of these proposed sites, high-resolution geologic and topographic maps are available.

Role of Humans in Intensive Mars Exploration

The intensive exploration of Mars envisaged here will require substantial technical capabilities on the Mars surface. Long paths must be traversed and explored geologically in detail. The scientific goals require samples to be collected, along with preliminary in situ analysis. To meet the sampling requirements, holes must be bored and drill cores extracted. Depths of up to 2 km would be desirable near the poles. The network of seismic and meteorological stations will require maintenance. Achieving a planet-wide understanding of the diverse provinces and processes will require extensive exploration over many parts of the surface. Although autonomous robot vehicles may perform many of the initial exploratory activities, the later, more comprehensive exploration will require capabilities that now are possessed only by humans.

The detailed exploration of Mars and its comparison to Earth, which is the objective of the proposed campaign to understand Mars, can be likened to the Apollo exploration of the Moon. The lunar astronauts successfully landed spacecraft in difficult terrain, selected a wide variety of samples, used simple tools to enhance sampling, and applied ingenuity and strength to overcome operational problems. These technical achievements were accomplished in relatively short stays. The crews traveled up to 25 km at speeds up to 10 km per hour in rovers, and selected and documented a wide variety of samples. Further, the crews have contributed to post-mission data analysis extending over the past 15 years.

The ability to use simple tools can probably be reproduced by autonomous vehicles. Time pressures would be considerably less for Mars surface exploration: robotic devices would have substantially more time to carry out these tasks. However, the intelligent and interactive selection of appropriate samples, and the concurrent and later provision of contextual information about them is well beyond the capability of current automated systems.

Mars shows a much more complex variety of geologic processes than the Moon. Since the martian surface is also weathered, discernible differences based on mineralogy, texture, and resistance to erosion may be important to interpretation. This greater diversity requires subtle judgments and interactions among observation, analysis, and interpretation. The total context of any sample will be very important in its scientific interpretation.

Two conclusions follow from these considerations. First, geologists, properly equipped, learning by observation, improvising as necessary, and advised from Earth, can be fast and effective at exploring the surface of Mars. Second, it is difficult for autonomous devices, even remotely linked to Earth by cameras, to achieve the scientific goals set forward above. Ultimately, to resolve the important questions and to compare Mars in detail with the Earth will require exploration capabilities on the surface of Mars possessed now only by humans.

A Phased Approach

The task group recommends an intensive study of Mars to be implemented by phases that begin with currently envisioned missions, progress to newly developed robotic and propulsion systems later, and culminate in manned missions to Mars. The Mars

Observer and Mars Aeronomy Observer are missions central to the broad program of solar system exploration. They will provide the global overview from which to mount and target later efforts at the Mars surface. An intensive Mars campaign would begin with a suite of unmanned missions that would establish surface measurement networks and return geologic samples that had been selected and gathered by automated techniques. We know from earth field studies and from the Apollo experience that the scientific utility of these returned samples will be limited by lack of information about their context. The intensive study of Mars must ultimately be extended to incorporate certain capabilities that currently only humans can supply: human intelligence is needed for reasoned sample selection, to provide contextual documentation, and to assist in the interpretation of sample analyses. In the meantime we should develop robotics and artificial intelligence, but these developments will not obviate the requirement for a human presence on Mars' surface to support the later phases of the Mars intensive study.

RECOMMENDATIONS

The goals of planetary exploration are achieved primarily through the analysis of data returned from spacecraft missions. Complementary remote observations are obtained using telescopes in the vicinity of Earth. Physical information on planetary materials is acquired from terrestrial prototypes, laboratory investigations, and earth orbital observations. In addition, theoretical modeling furthers scientific understanding. A viable planetary program must contain all five elements—spacecraft missions, telescope observations, field investigations, laboratories, and models—and the task group's recommendations speak to each.

Exploration of the Solar System

Figures 2.2, 2.3, and 2.4 in Chapter 2 show the status of planetary exploration expected by 1995, and the missions recommended for the period 1995 to 2015. The scientific objectives of these missions are outlined in Table 2.1. It must be emphasized that the near-term and the far-term planetary exploration projects are proposed in a logical order, following the sequence of reconnaissance, exploration, and intensive study. However, no priorities are implied by the ordering. Thus, if the status of planetary studies in 1995 is not as projected, then those studies that have not been completed should still have priority over the longer-term projects.

The proposed missions follow the phased approach for solar system study. By 1995, the exploratory and reconnaissance phases should have been completed for the inner planets with the important exception of Mercury. This planet lies deep in the Sun's gravitational well, so that it is difficult for spacecraft to reach, particularly with approach velocities small enough to allow capture into an orbit. Nevertheless, the unique properties of Mercury, especially the high mean density, which is probably indicative of strange internal chemistry and mineralogy, make its study of the greatest interest for comparative planetology. A Mercury orbiter mission will be needed to complete the reconnaissance stage of that planet's study.

The other proposed inner planet mission concepts involve the intensive study phase. For the Moon, the near-term geophysical studies from orbit will undoubtedly raise questions requiring additional sample return for their resolution. Some of these samples will be needed from the far side of the Moon, and one contemplated project would select and return such samples. Complementary to the sample return will be global geophysical studies of the planet, for which the establishment of a network of stations will be required. In due course we expect that there will be numerous practical reasons for establishing a permanent base of operations upon the Moon, and the course of lunar research outlined here will prepare the way for such a manned lunar base.

For Mars, the stage of intensive study began with the Viking mission and will continue with the Phobos mission. The next major steps, which are of the greatest scientific importance, will be Mars rovers, the establishment of a global sensor network, and the selection and return of samples for analysis in terrestrial laboratories. When the sphere of human habitation in space enlarges to encompass Mars, the establishment of a manned base there will become desirable, and the studies outlined here will be necessary precursors.

In the case of Venus, a good map is partially in hand; completion is expected with the planned radar mapper mission (Magellan). Current lack of this map inhibits detailed projections for future missions. An initial set of geochemical and mapping information has been obtained from Soviet investigations. The hostile environment of the planet requires much more technological development for future missions than is the case for the other terrestrial planets. Nevertheless, the kind of geophysical and geochemical information desired from Venus is similar to that desired from the other terrestrial planets, and the means needed to acquire this will include probes, the establishment of a global network, and sample returns. Accomplishing these objectives will provide interesting technological challenges.

Detailed planning for the intensive study of Mercury must await the imaging of the unseen hemisphere and the geophysical and geochemical mapping that will be done in an orbiter mission. With those data in hand it will be possible to plan the kind of surface investigations a lander will allow and to plan the return of samples for laboratory analysis. It is desirable that these follow-on missions should be done within the contemplated time period of this study.

For the four giant planets of the outer solar system the reconnaissance stage of planetary study requires an orbiting spacecraft and an atmospheric entry probe. By the 1995 to 2015 period this should have been accomplished for Jupiter by the Galileo mission. Orbiter and probe missions for Saturn, Uranus, and Neptune are then of the highest priority for the outer solar system. In the case of Jupiter, a different kind of orbiter—one in a polar orbit—is needed as a follow-on to Galileo in order to study the rich inner magnetosphere of the planet in detail.

The satellites of the giant planets are of great interest, especially the larger ones. The Galileo mission is expected to complete the reconnaissance of the Galilean satellites of Jupiter. Titan, the large satellite of Saturn, has a substantial obscuring atmosphere and thus should have a dedicated orbiter-probe mission of its own. The Saturn orbiter, after delivery of its probe, may become dedicated to studies of the rings, and thus would be unavailable for reconnaissance of the more distant regular satellites. In this case, a separate orbiter would be desirable to examine these satellites.

The system of Pluto, consisting of an icy planet of low mass with a satellite of high relative mass, is of great intrinsic interest. By 1995 it will not have undergone exploration from space. It is therefore recommended that, owing to the very long flight time to the planet, the exploration and reconnaissance stages of study be combined, and performed by a dedicated orbiter. Extensive earth-based observations of the system, using the Space Telescope and other major instruments, will be needed to plan this mission.

Once the outer planet missions outlined above have been carried out, the stage of intensive study can begin. For the gas giant planets deep probes will be needed in order to study the atmosphere to greater depths and to refine the measurements made with the first generation of entry probes. Such deep probes are recommended for Jupiter, Saturn, and Uranus by 2015.

Intensive study of the satellites of the giant planets will require surface investigations by means of landers, which should be able to emplace networks. It is recommended that several selected satellites of the Jupiter and Saturn systems be investigated in this way by 2015. A lander of special design will be needed for Titan in view of its atmosphere; such a lander mission should be a rich source of information both about the atmosphere and whatever kind of surface exists, even if covered by liquid, but the design of such a lander must await results from the Titan orbiter-probe.

The reconnaissance of comets should have commenced by 1995 with a comet rendezvous mission. The material constituting comets is likely to be the most primitive form of matter preserved from the environment of the early solar system that will ever be available to us. Thus sample return is vitally important for further studies of comets. It is recommended that a fragmented sample be returned by means of a fast comet flyby spacecraft that can easily return to the Earth. This should be followed by a more advanced mission involving a rendezvous in which special care will be taken to obtain and maintain several samples of the comet in their pristine form, and to continue to maintain them in that state during the return to the vicinity of the Earth.

It is likely that a small number of asteroids will have been examined in a flyby mode by 1995. However, these bodies are highly diverse in their properties, and the early glimpses should be followed by a multiple rendezvous mission, in which a variety of surface analyses can be made, including some interactive analyses. These would pave the way for a later mission that would return samples from asteroids.

The techniques required for the search for other planetary systems will continually improve. The task group recommends a dedicated telescope associated with the space station for this purpose. Even larger facilities, perhaps space-based and perhaps lunar-based, will be valuable for this purpose as well as for many other uses that require high-resolution imaging of the solar system and beyond.

Earth-based studies are envisioned as a continuous program throughout the 1995 to 2015 period. Included are observations from Earth and near-Earth orbit of solar system objects and the search for other planetary systems, laboratory experiments related to planetary processes, and analyses of meteorites.

Most of the missions shown in Figures 2.3 and 2.4 can be achieved with existing or near-term technology. The study of Uranus, Neptune, and many of the small bodies will require the development of low-thrust propulsion. The intensive study of Venus cannot begin without extending the high-temperature survival of electronics; in addition, the return of Venus samples will require significant developments in propulsion. The task group recommends that development efforts in these areas of technology be initiated as soon as possible.

Telescopes on Earth and in earth orbit complement space probes by providing observational data that, while usually of lower spatial resolution, can be synoptic in scope and quickly responsive to phenomena. The task group recommends support of a vital program of planetary astronomy and particularly encourages preparations to use new generations of astronomical telescopes for planetary observations.

An integral part of any mission is adequate support for analysis and interpretation of the data returned from it. Studies in the field, in earth orbit, and in laboratories are required to provide corroborating "ground truth," calibrations, and fundamental physical constants. The task group recommends continual upgrading of laboratory instrumentation and of computing equipment used for data analysis and theoretical modeling.

The missions and activities outlined here would address the objectives of solar system exploration outlined earlier. We have learned a great deal in the past 20 years, and the next 30 can be even more productive. We can also expect to support the extension of the sphere of human habitation into space by improving our knowledge of planetary environments. The task group recommends that the program be implemented as vigorously as allowed by economic and national goals.

Exploration of Mars

The task group strongly recommends a continuing scientific exploration of the entire solar system, primarily using instruments on automated spacecraft. There is much to be done to complete even a first-order look at our planetary neighborhood. That investigation must continue. But a major Mars campaign could be carried out concurrently, in the same way that the first decade of planetary exploration was carried out by Mariner spacecraft in parallel with the Apollo Moon program.

Mars is the most Earth-like of the other planets, displaying the full range of terrestrial phenomena (except, possibly, life), although frequently in greatly modified form. Mars is an ideal location for the study of a long geological history parallel to that of the Earth, including both volcanic and tectonic activity; for examining the chemical evolution of an atmosphere and its interaction with a surface, for investigating a complex meteorology including cyclic transport of volatiles between surface, atmosphere, and polar caps; and for exploring evidence of climatic cycles. The presence of channels of a variety of ages indicates episodic flow of large amounts of water. Mars may also have preserved evidence of prebiotic chemical evolution, or even possibly of the development and evolution of an indigenous biota.

In the extensive exploration of Mars envisaged here, humans will play an essential role. Thousands of kilometers of the surface will need to be traversed and explored geologically in detail. Samples must be collected and given preliminary in situ analysis, holes bored and cores extracted, and automatic stations to monitor meteorological activity, seismicity, and heat flow emplaced and maintained. In some areas, particularly on and near the polar caps, extensive drilling and perhaps excavation should be undertaken. In order to reduce costs and increase the efficiency of the operation, equipment for the manufacture of rocket fuel and of essential water and oxygen for personnel must be established. It is difficult to imagine that such an extensive, detailed, planet-wide exploration program could be carried out effectively by autonomous robot vehicles alone. The direct application of human knowledge and ingenuity to the detailed exploration of a new world is likely to lead to the maximum return and the deepest level of understanding.